Springer Natural Hazards

The Springer Natural Hazards series seeks to publish a broad portfolio of scientific books, aiming at researchers, students, and everyone interested in Natural Hazard research. The series includes peer-reviewed monographs, edited volumes, textbooks, and conference proceedings. It covers all categories of hazards such as atmospheric/climatological/oceanographic hazards, storms, tsunamis, floods, avalanches, landslides, erosion, earthquakes, volcanoes, and welcomes book proposals on topics like risk assessment, risk management, and mitigation of hazards, and related subjects.

More information about this series at http://www.springernature.com/series/10179

Mary Ann Cooper • Ronald L. Holle

Reducing Lightning Injuries Worldwide

Springer

Mary Ann Cooper
African Centres for Lightning and Electromagnetics Network Inc. (ACLENet)
River Forest, Illinois, USA

Professor Emerita,
University of Illinois
Chicago, USA

Ronald L. Holle
Holle Meteorology and Photography
Oro Valley, AZ, USA

ISSN 2365-0656 ISSN 2365-0664 (electronic)
Springer Natural Hazards
ISBN 978-3-030-08480-6 ISBN 978-3-319-77563-0 (eBook)
https://doi.org/10.1007/978-3-319-77563-0

Printed on acid-free paper

This Springer imprint is published by the registered company Springer International Publishing AG part of Springer Nature.
The registered company address is: Gewerbestrasse 11, 6330 Cham, Switzerland

Foreword

Lightning has intrigued mankind over the centuries, yet it is still one of the least understood natural phenomena commonly observed by the public. It strikes somewhat randomly, kills or injures many thousands of people worldwide each year, and causes billions of dollars in damages. Understanding this powerful and dangerous phenomenon and its impacts on people's lives is an important step in reducing deaths, injuries, and damages from lightning around the world.

Both Dr. Mary Ann Cooper and Ron Holle have been leaders in the effort to help people understand the dangers of lightning, its impacts, and what people can do to protect themselves and their property. Over the years, they have accumulated a wealth of knowledge and have shared their knowledge with the medical and scientific communities through their writings and presentations. Both have also been at the forefront in developing lightning safety guidelines.

Dr. Cooper has been a leader in the medical community in investigating, understanding, and documenting the short- and long-term effects of a lightning strike on the human body. She has worked with numerous lightning strike survivors and their families to not only understand the physical injuries that lightning causes, but also the mental, psychological, and financial effects on the victims and their families. She has been a resource for doctors around the world to help them understand the medical impacts of lightning injury and has shared her knowledge to help in the treatment of lightning strike survivors.

Ron Holle has been a leader in investigating and documenting lightning deaths and injuries worldwide. He has documented and analyzed US lightning deaths and injuries for more than 30 years. In addition, he has investigated historic US lightning fatality data to determine differences between recent fatalities and those that occurred more than 100 years ago when the United States would be considered a developing country by today's standards. This has allowed him to understand the demographics of victims and the situations that put people at risk, both now and when the population of the United States was more rural. In addition, his work with global lightning detection systems has allowed him to understand the distribution of lightning around the world and identify areas where large populations are most vulnerable.

I have personally known Mary Ann and Ron for almost two decades as part of the National Oceanic and Atmospheric Administration's (NOAA's) Lightning Safety Team and through our joint efforts to reduce lightning deaths and injuries. During this time, their knowledge and dedication to the lightning safety effort has been critical to the success of NOAA's lightning safety campaign. Both have contributed their professional expertise and personal time to help make the information on NOAA's lightning safety website the best in the world.

While NOAA's efforts have focused mainly on safety issues in the United States, Dr. Cooper and Ron Holle have expanded their personal efforts to other areas of the world, and, in particular, the unique challenges of the developing world. Most recently, they have been working together with leaders of developing countries to understand the specific challenges that those countries face and to develop ideas of what could be done to address those challenges. In some ways, the challenges that those countries face are similar to the challenges faced by the United States and other developed countries more than 100 years ago. Personally, I look back to the 1950s and 1960s when I was growing up in rural Pennsylvania. At the time, there were many small farms and people generally waited for it to start raining before going inside. As a result, the US lightning death toll was typically between 100 and 200 people per year. In fact, in the early 1940s, the United States typically saw between 300 and 400 lightning fatalities a year. While many things in the United States have changed since then, I truly believe that a better understanding of the dangers of lightning and improvements in the medical treatment of victims have both contributed greatly to the lower US lightning death toll.

With recent advances in technology, telecommunications, and reporting, governments of developing countries now have become more aware of the large numbers of people killed by lightning each year in their respective countries. This information has led to calls for greater efforts to minimize the threat to vulnerable populations. While education is a key component in reducing deaths in those countries, there are other critical issues such as the need for structures that provide safety.

In *Reducing Lightning Injuries Worldwide*, the authors have put together a comprehensive background on lightning and the medical effects on the human body; issues related to lightning safety and lightning protection; and documentation on worldwide lightning fatalities. The book also discusses the differences between developed and developing countries and the challenges that the developing countries face in trying to reduce lightning fatalities. Finally, and most importantly, this book offers suggestions and recommendations for reducing global lightning fatalities, based partially on the efforts that have worked in developed countries, but with consideration given to the limited resources available in developing countries and an understanding of cultural background differences.

For all those who are interested in protecting people from the potentially devastating impacts of lightning, *Reducing Lightning Injuries Worldwide* will provide valuable information and ideas to help reduce lightning deaths and injuries across the globe.

John Jensenius
Lightning Safety Specialist
National Weather Service, NOAA
Gray, Maine, USA

Preface

Why Is This Book Needed?

Ideally, we should be able to prevent all lightning injuries, deaths, and property damage. Prevention is always better than burying the dead, taking care of those who survive, or trying to repair the damage to property and electronics. However, we have not reached the point where all injuries and property damage can be prevented.

Ideally, for the lightning injuries that we cannot prevent, physicians should know how to treat it based on known and certain pathophysiology using research based therapies. The pathophysiology of lightning injury may never be known because of the difficulty in doing research with lightning on living tissue – or even on nonliving tissue. To be brutally honest, papers on the pathophysiology of lightning injury have almost always been more educated speculation than fact supported by research.

Therefore, the REASON THIS BOOK IS NEEDED is because PREVENTION is better than caring for the survivors of lightning injury.

Reason for This Book (and How to Use It)

The authors were solicited to write a book about the current state of lightning studies. We proposed to write a book that gives an overview of the current state of knowledge of many aspects of lightning, some of which are more within our particular areas of expertise than other areas. We see each chapter standing independently but with references to other chapters that may be pertinent to questions that are more specific. We have been privileged to work together for nearly three decades on lightning injury prevention, the ultimate goal for this book, and to have known and often worked with the giants in lightning over that time.

The book is intended to provide a resource to understand the current situation in the developed and lesser-developed areas of the world and how to address reducing lightning casualties in vulnerable areas of the world. How, when, where, and during what types of activities people become lightning casualties will be addressed, as well as a description of the distribution of lightning around the world and the factors responsible for its occurrence.

Information on lightning fatalities and injuries cuts across many disciplines: public health, medicine, trauma studies, pain management, injury prevention, electrical engineering, physics, architecture and structural protection, geography, commerce, business, education, communications and media, psychology, neuropsychology, as well as social science including what individuals and populations believe and how they respond to the threat, mining, utility management, aviation, and many, many more (Andrews 1995).

Depending on the discipline of the researcher, lightning fatality reduction is often considered in isolated efforts that do not provide complete solutions; nevertheless, the possibility exists to reduce the large loss of life due to lightning in the developing world. To address these issues, we intend the audience for the book to be those in public health, public policy programs, government and private organizations involved in improving public safety in the developing world, and others who want to address the threat of lightning and decrease injuries and deaths.

This book is also meant to serve as a resource and sometimes a starting point for students and faculty who may be interested in initiating local studies and projects related to reducing global lightning casualties or in pursuing more in-depth studies, whether an undergraduate project, master's thesis, or doctoral dissertation. Each chapter serves as an introduction, not an exhaustive discussion – much more exhaustive books are available for many of the topics. Each chapter will include questions that naturally arise from curiosity about lightning and the different areas each chapter covers. Many of these have not been answered and could serve as a basis for forming more specific research questions. Some of the questions will be for the reader's particular situation and encourage critical thinking.

Each chapter ends with a list of some key references that can serve as a beginning reading list for the student. We have also cataloged lightning injury prevention program activists and researchers so that students, public health policy makers, and others may access their work or contact them for more information, collaboration, or mentoring for programs they wish to start.

With very few exceptions, we have found the lightning injury prevention community worldwide to be warm, sharing, concerned people who do most of their work on their own time and often on their own money, hoping to prevent injuries and save lives. For those of you who wish to join us, we welcome you and will do what we can to help you!

Reference

Andrews CJ (1995) Keraunomedicine – a discipline come of age. Ann Emerg Med 25(4):543–545

River Forest, IL, USA — Mary Ann Cooper
Oro Valley, AZ, USA — Ronald L. Holle

List of Abbreviations and Definitions

Lightning

CG flash	Cloud-to-ground flash. A cloud-to-ground lightning flash has one or more return strokes.
CG return stroke	Cloud-to-ground stroke. One of the components of a cloud-to-ground flash. A flash has one or more return strokes, averaging four to five strokes per flash.
Flash	The entire sequence of a lightning event, starting with its initiation in the cloud through the last portion of its visible light.
GLD360	Global Lightning Dataset lightning detection network.
IC	In-cloud lightning. About three times as many lightning events occur in cloud without reaching the ground as those that contact the surface of the earth.
Hardening	Making equipment or infrastructure lightning resistant.
NLDN	National Lightning Detection Network.
LCC	Long continuing (continuous) current occurs when current continues to flow between the individual cloud-to-ground strokes of a cloud-to-ground flash.
Lightning density	The number of lightning events per area per time, such as CG flashes per square kilometer per year.
"Lightning safe" areas	Only two areas are identified as "lightning safe": substantial buildings, defined as those having plumbing, wiring, and metal structural elements in the walls, and fully enclosed all metal vehicles.
Substantial building	A structure that is safe from lightning with paths for lightning to follow through grounded wiring and plumbing in the walls, and may have metal structural members (Chap. 16).
Total lightning	Sum of CG and IC data.

Meteorology (American Meteorological Society, 2015)

30–30 rule	The first 30 refers to the number of seconds between seeing lightning and hearing its associated thunder; a 30-second interval refers to six miles (10 km). The second 30 refers to the number of minutes to wait until resuming outdoor activity after the last lightning or thunder within a specific range.
Convection	Primarily vertical cloud development.
Tropical cyclone	An organized low-pressure system over warm oceans. A tropical depression has winds up to 17 m s^{-1} (34 knots), a tropical storm has winds of 18–32 m s^{-1} (35–64 knots), and a severe tropical cyclone (also called hurricane or typhoon) has winds of 33 m s^{-1} (65 knots) or more.
Derecho	An organized area of convection with widespread strong winds in the evening and nighttime, primarily during the middle-latitude summer.
Diurnal	Daily variations that occur within a 24-h cycle.
Equatorial trough	Nearly continuous belt of low pressure between the subtropical high-pressure belts of the Northern and Southern hemispheres. It moves into or toward the summer hemisphere. Also called the ITCZ.
Flash-to-bang	The time interval in seconds between seeing lightning and hearing its associated thunder.
Inter-tropical Convergence Zone (ITCZ)	A shallow low-pressure zone around the *equator*, where winds tend to converge from both hemispheres. Also called the equatorial trough.
Large-scale systems	Meteorological systems with horizontal scales of thousands of kilometers.
Mesoscale	Atmospheric phenomena with horizontal scales ranging from a few to

	several hundred kilometers, such as thunderstorms.
Mesoscale convective systems (MCS)	Prolific lightning producers that cover very large areas, usually over land, are strongest at night, and last up to 18 h (Sect. 13.2).
Middle latitudes	Between 23.5 and 66.5 North and South latitudes where there are often four distinct seasons and weather systems frequently travel from west to east.
Monsoon	A seasonally reversing wind accompanied by corresponding changes in precipitation.
Small-scale systems	Meteorological systems that are tens of kilometers across.
Subsidence	Descending motion of air in the atmosphere.
Tropical	Between 23.5 North and South latitudes.
Turbulence	Random, continuously varying air motions in addition to the broader-scale air motions.
Updraft	Upward motion within cumulus convection.
Westerlies	Prevailing direction of motion of weather systems in the middle latitudes between 23.5 and 66.5 North and South latitudes.

Human Impacts

Cardiac	Having to do with the heart.
Casualty	The sum of deaths and injuries.
Cognitive	Having to do with thought or thought processing including memory, perception, executive function, and other brain functions. Usually does not include motor function.
Cranial	Having to do with the head (cranium).
Death, fatality	A person killed by lightning.
Dysesthesia	Abnormal (dys), usually unpleasant, sensory perception (esthesia), which may be called numbness, burning, tingling, shooting, itching, painful, or other adjectives the person chooses to use.

Hyperacusis	Sensitivity to noise.
Hypertension	High blood pressure.
Injury	A person injured by lightning, including both those killed and those who survive.
Injury cascade	The order of normally expected bodily responses after an injury.
Keraunoparalysis	Usually temporary paralysis of legs and/or arms that lasts for several minutes caused by lightning (kerauno).
Neurologic	Having to do with the brain, spinal cord, autonomic nervous system, or nerves.
Neuropathy	Pathologic (abnormal) signaling from injured/healed nerves: most commonly characterized by the person as painful.
Orifice	Normal opening on a body such as the mouth, nostrils, or anus.
Paresthesia	Abnormal sensory perception, most commonly numbness, but may also be called shooting, burning, tingling, itching, painful, or other adjectives that the person chooses to use.
Physiology	The branch of biology that deals with the normal functions of living organisms and their parts.
Pathophysiology	Physiology of an abnormal state, usually caused by an injury, illness, or toxin.
Photophobia	Sensitivity to light
Post-concussion syndrome	Set of signs and symptoms of brain injury caused by a fall, explosion, or other concussive injury. Common symptoms include headache, dizziness, fatigue, irritability, anxiety, insomnia, loss of concentration and memory, ringing in the ears, blurry vision, and noise and light sensitivity.
Pulmonary	Having to do with the lungs.
Sequela (plural sequelae)	After-effect or complication after the initial (acute) phase of an injury or illness.
Thrombosis	Blockage of a blood vessel by a clot.
Trauma	The exposure to some source of energy (mechanical, electrical, thermal, radiation, or chemical) in an intensity exceeding the tolerance level of the host (Chap. 2) (Navarrete-Aldana 2016).
Tympanic membrane (TM)	Ear drum.

Organizations

ACLENet African Centres for Lightning and Electromagnetics. A pan-African network of national and regional centers dedicated to decreasing deaths, injuries, and property damage from lightning (Chap. 18) (ACLENet.org/).

LSESSI Lightning Strike and Electric Survivors, International. A nonprofit organization in the United States dedicated to survivors, their families, and other interested parties (Cooper and Marshburn 2005; Chap. 3) (www.lightning-strike.org/).

References

American Meteorological Society (2015) Glossary of meteorology. http://glossary.ametsoc.org/

Cooper MA, Marshburn S (2005) Lightning Strike and Electric Shock Survivors, International. NeuroRehabilitation 20:43–47

Navarrete-Aldana N (2016) Lightning injuries in Colombia. From lack of awareness to prevention. Paper presented at the world meeting on lightning, Cartagena de Indias, 6–8 Apr 2016

Acknowledgments

The authors of this book have been inspired by a host of collaborators and colleagues in studies of the impacts of lightning on people in the last three decades. No list is complete, and we emphasize that those not included should realize it was an oversight.

From the medical community, a group of medical doctors is highly relevant. These are Dr. Chris Andrews in Australia, Dr. Ryan Blumenthal in South Africa, Dr. Michael Cherington and Dr. Phil Yarnell in Colorado (United States), and Dr. Norberto Navarrete-Aldana in Colombia.

From the United States meteorological community, Dr. Raul López of the National Oceanic and Atmospheric Administration (NOAA) research laboratories participated in many studies. Within the National Weather Service of the National Oceanic and Atmospheric Administration, Mr. John Jensenius, Ms. Donna Franklin, Mr. Steve Hodanish, and Mr. Charles Woodrum continue to be active. Mr. William Roeder in Florida and Mr. Ken Langford in Colorado are also frequent collaborators.

From the lightning physics and protection community, Dr. Vladimir Rakov in Florida (United States) and Dr. E. Philip Krider, Dr. Kenneth Cummins, Dr. Amitabh Nag, and Ms. Daile Zhang in Arizona (United States) have often been sources of support and advice.

Additional champions of lightning safety include Dr. Katie Walsh Flanagan in North Carolina (United States), Dr. Chandima Gomes and Dr. Mohd. Zainal ab Kadir in Malaysia, Dr. John Gookin in Wyoming (United States), Mr. Richard Tushemereirwe in Uganda, Mrs. Foster Chileshe Lubasi in Zambia, Mr. Daniel Esteban Villamil Sierra in Colombia, Mr. Munir Ahmad in Bangladesh, and Dr. Shri Ram Sharma in Nepal.

Contents

Part I
How Lightning Kills, Injures, and Causes Damage

Chapter 1
Introduction

Abstract Lightning causes injuries and deaths in nearly all parts of the world, more commonly in tropical and subtropical areas than in middle latitudes and rarely in the arctic areas. The distribution of injuries is partly due to lightning density, the number of times lightning hits a particular area over a particular time space, but also to population density and risk of exposure. This book will explore those features.

1.1 Problem Statement

Lightning causes injuries and deaths in nearly all parts of the world, more commonly in tropical and subtropical areas than in middle latitudes and rarely in the arctic areas. The distribution of injuries is partly due to lightning density, the number of times lightning hits a particular area over a particular time space, but also to population density and risk of exposure. This book will explore those features.

While lightning injuries occur all over the world, they have generally been poorly documented except in a few countries. Natural hazards such as hurricanes, tornadoes, floods, and extreme cold or heat tend to kill more people in a particular incident and are more likely to involve government response, an outpouring of aid to the victims, data collection, and attempts at prevention. They are also more likely to be published in the media, making the public more cognizant of these risks.

Thunderstorms tend to be small, often only a few kilometers in size, and form and disperse rapidly compared to hurricanes and floods. They are everyday, commonplace events and seldom newsworthy by themselves. Similarly, lightning injuries, with some exceptions, tend to injure only one or two individuals at a time, also making them less likely to come to the attention of the media or government, especially in rural or less developed areas where communication systems may be poor or nonexistent. Survivors may not seek medical care to be entered into a data system. Nevertheless, lightning kills and injures a significant number of people ever year as well as livestock, often the measure of wealth in developing nations, and damages property including infrastructure in industries such as utilities, communications systems, electronics, and many others, adversely affecting not only the company but also communities and nations struggling to develop stable economies.

M. A. Cooper, R. L. Holle, *Reducing Lightning Injuries Worldwide*,
Springer Natural Hazards, https://doi.org/10.1007/978-3-319-77563-0_1

The number of lightning fatalities and injuries in the developed world has been steadily declining over the last century. However, the number of lightning casualties in the lesser developed regions is not decreasing due to billions of people living in lightning-vulnerable housing, working outdoors in labor-intensive manual agriculture, and other identifiable socioeconomic factors. The global lightning impacts to people may be as high as 24,000 deaths and 240,000 injuries per year.

Chapter 2
Mechanisms of Lightning Injury

Abstract The only mechanism of lightning injury that most people consider is the direct strike, from cloud to ground. In actuality, there are five mechanisms of injury by which a person can be impacted by the electrical nature of lightning. In addition, there are several possible secondary mechanisms of traumatic injury. This chapter will describe all of these mechanisms and discuss the differences in developed versus developing countries.

2.1 Introduction

Lightning can cause devastating injuries, but generally not in the way that most people would imagine nor conclude from their knowledge of electricity and lightning. Initially, if one asked a person on the street what types of injuries lightning would cause, they would probably say that it would be a serious burn injury. Some might add that it could cause the heart or other "electrical" organs to fail. A very few might think of thunder and explain it as a concussive or explosive injury. Nearly everyone speaks of lightning as if all injuries are from direct hits, mostly because most people don't know about the other mechanisms of injury (Cooper and Holle 2010).

Lightning injury is a traumatic injury. One does not have to be a physician to appreciate the injury, disability, and death caused by lightning nor to recognize the need for lightning injury prevention.

Definition of Trauma
The exposure to some source of energy (mechanical, electrical, thermal, radiation, or chemical) in an intensity exceeding the tolerance level of the host (Navarrete-Aldana 2016).

M. A. Cooper, R. L. Holle, *Reducing Lightning Injuries Worldwide*,
Springer Natural Hazards, https://doi.org/10.1007/978-3-319-77563-0_2

2.2 Mechanisms of Lightning Injury

2.2.1 *Electrical Injury by Lightning*

As an electrical force, lightning can cause injury through five primary mechanisms. Their estimated frequency is shown in Fig. 2.1, and illustrations of each are in Fig. 2.2 (Cooper et al. 2008):

1. *Direct strike*: There is nothing between the person and the lightning that contacts (or "attaches to") them. In developed countries, it is estimated that 3–5% of fatalities are caused by direct strike. Although it is thought that direct strike is more likely to be fatal, there is no clinical nor experimental data to support this hypothesis. It is not known whether the prevalence of death and injury by direct strike for developing countries is the same (Cooper 2012).
2. *Contact voltage*: Lightning hits something else first and travels through that pathway to affect someone who is holding onto the energy transmitter. Examples are someone turning on a water faucet when lightning has hit the ground a distance away and been transmitted through the water or plumbing or someone talking on a hardwired phone (Andrews 1992; Andrews and Darveniza 1989). It is estimated that contact injury causes approximately 15–25% of deaths in developed countries (Fig. 2.1).
3. *Sideflash or splash*: Lightning hits another object, and a portion of the energy jumps to a nearby person to complete its path to ground. An example is someone standing under a tree. About 20–30% of lightning deaths in developed countries are caused by sideflash.
4. *Ground current* (also called *step voltage, ground potential,* and a number of other names): Lightning hits the ground a distance away from a person and spreads through the ground nearly radially. It may go up one leg and down the other in a standing person or from head to foot in someone sleeping in a tent. It causes 40–50% of deaths in developed countries. Ground current can be divided into two subsets, both of which may occur with any strike:

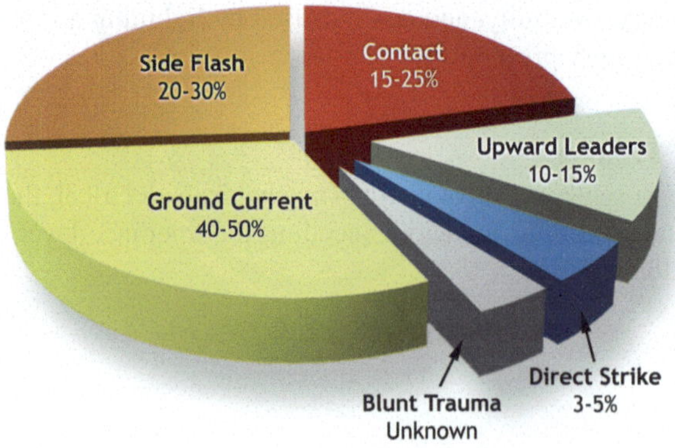

Fig. 2.1 Mechanisms of lightning injury. (Courtesy of Environment and Climate Change Canada (Cooper et al. 2008))

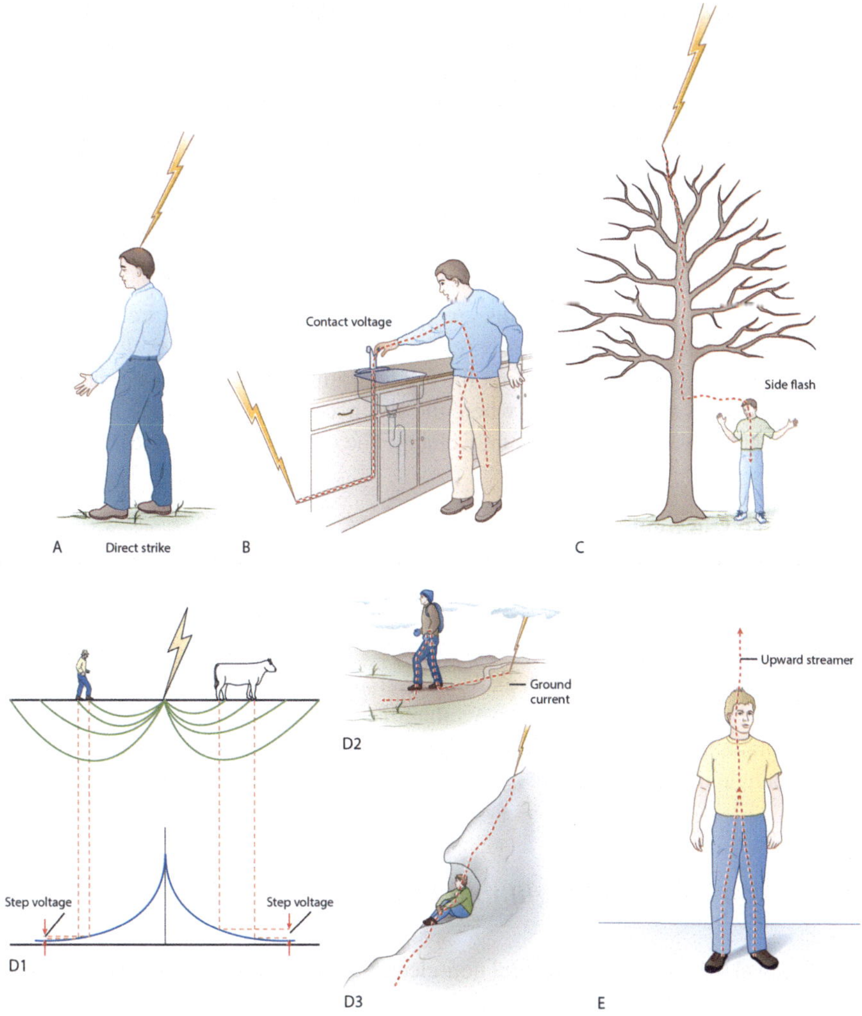

Fig. 2.2 Mechanisms of lightning injury. (**a**) Direct. (**b**) Contact. (**c**) Sidesplash/flash. (**d1**) As lightning energy spreads out from the strike point, the energy decreases. A potential difference can be generated between a part of the body is closer to the strike and that further away, setting up a current through the body. (**d2**) Step voltage or ground current traveling through the ground. (**d3**) Ground arcing across the mouth of a cave. (**e**) Upward streamer. (Cooper et al. 2017 with permission of Elsevier)

(a) *Step voltage* occurs when the energy stays in the ground as it contacts the person and returns to the ground from the person. This tends to deliver lower energy to the person.

(b) *Ground arcing* occurs when the energy jumps through the air, such as across the mouth of a cave where a person is sheltering. Because the arc must have energy high enough to overcome the dielectric constant of air, ground arcing generally involve a considerably higher energy and temperature.

5. *Upward streamer (upward leader)*: Any electric field, such as in a thunderstorm, will induce an opposite charge in objects close to it, including a television tower, tree, person, or blade of grass (Cummins et al. 2018). These charges can actually coalesce to form an upward leader from the object toward the strong electric field. Even if this streamer does not attach to the downward leader of lightning from the overheard thunderstorm, it has enough energy to injure a person (Anderson 2001; Anderson et al. 2002; Cooper 2002; Anderson and Carte 2009). Upward streamers are thought to cause about 10–15% of lightning deaths in developed countries.

As noted, these five mechanisms are primarily electrical in nature. While the distribution of deaths by lightning is reasonably well known for developed countries, the distribution of mechanisms for nonfatal injuries is not and neither is the distribution in developing nations where the absence of safer areas such as substantial housing and all metal vehicles may change the exposure (Cooper 2012).

Research Questions

Is this distribution correct? What is the distribution of mechanisms for nonfatal lightning injury? What is the distribution of mechanisms in developing countries for either deaths or injuries? Is the distribution of mechanisms the same in developing countries as developed countries? Is the distribution dependent on geography, such as forests versus deserts, or the availability of safer shelters? Is there a predictive value that can be assigned to a particular mechanism that would be useful in anticipating the level of injury or useful in treatment and rehabilitation?

2.2.2 Non-Electrical Lightning Injury

People can also be injured by several non-electrical mechanisms in the case of lightning, as in the following three methods:

1. *Barotrauma* (concussion, blast, blunt force, or explosive injury): A person is close enough to the lightning channel to experience a rapid outward movement of air, similar to an explosion that may be enough to knock them off their feet or cause concussive injury to internal organs. This has been described for at least a century but recently investigated more thoroughly by a forensic pathologist who calculated the pressure wave from lightning at a distance of 10 meters from the strike to be similar to the force of a 5 kg TNT bomb (Blumenthal and West 2015).
2. *Shrapnel or missile injury*: A person receives penetrating trauma as a lightning impact blasts shrapnel such as tree bark or material from a cement walkway into them (Blumenthal 2012). The explosion in this case is usually hypothesized to be from pockets of moisture in the material that expands as they are heated by lightning to cause a vapor explosion, blasting more superficial material outward from the main mass.

3. *Blunt injury*: A person is thrown a distance by lightning-induced muscular contraction. This can result in trauma similar to a fall, often with musculoskeletal injury.

Blast injury involves an explosive overpressure force hitting the person. Blunt injury occurs after the lightning strike and, similar to a fall or concussive injury, is a secondary force suffered after the strike where injuries may occur, including damage to internal organs as well as musculoskeletal injury. Both, along with shrapnel injuries, may occur to the same person. All may overlay or complicate any of the electrical mechanisms and cannot always be easily separated.

Research Questions

How would you test whether different mechanisms of injury are more likely to cause death or certain injuries than other mechanisms? How do these different mechanisms affect the "dose" of energy delivered to the person? Is the injury proportional to the "dose" (that is, can the "dose" be used to predict the severity of the injury)? Can blunt injury from being thrown be separated from injury by barotrauma? Is the brain injury suffered by many lightning survivors caused by barotrauma, by the electrical injury, by a humoral release, by rotational injury within the skull, by blunt force caused by being thrown a distance by muscle contraction or from some combination of these factors? How would you test your hypotheses?

2.2.3 *Burn Injury by Lightning: How It Is Different than Burns from Other Electrical Injuries*

While lightning is an electrical phenomenon, its characteristics are quite different from either household electricity or high-voltage electricity, both referred to from here on as "technical electricity" or simply electricity. Lightning has phenomenally higher voltage and usually amperage, but the time of exposure is only in the range of microseconds, unless the lightning exposure is from the relatively infrequent long continuing current (LCC), which still only lasts about a portion of a second (Saba et al. 2006; Rakov 2016). Cloud-to-ground lightning can lower either positive (10–20% of lightning) or negative (80–90%) charge to ground. Negative strokes are more common and, on average, of less magnitude and duration. LCC occurs more often in positive cloud-to-ground flashes than in negative lightning (Rakov 2016).

In addition, the electromagnetic waveshape of lightning is far different from regular alternating current or direct current. If one were to imagine the body as a tin can, then regular alternating current electricity would jump in and back, partially "filling" and "emptying" the tin can with every other reversal of current – and insulting the can (body) with each exposure. Additionally, the total exposure can be prolonged for several seconds, providing a large dose of traumatic energy, heating the tissue through which it is flowing, and creating massive burns, both internally and externally, particularly if it is from a high-voltage source. Even much lower-voltage household current, though it does not create impressive burns, can cause cardiac arrest and nerve damage.

With lightning, the rate of rise and the amount of current are so rapid that the tin can is "filled" almost instantaneously and the rest of the lightning spills over all around the can – a phenomenon appropriately named flashover. This takes place so rapidly that the body hardly experiences the lightning for more than a few microseconds at most. Despite the high voltage and amperage, the traumatic dose of electricity that lightning delivers to the body is actually very low. Note: This is not meant to be a technically quantifiable analogy but a visual thought aid in explaining the causality and injuries that are seen.

The lightning energy that flows around the body in flashover may cause secondary injuries as it turns sweat or rainwater on the body into steam (sometimes ripping apart the clothing or shoes from the steam vapor explosion), singes hair, heats metal objects such as coins, necklaces, or belt buckles, fuses low-melting point materials such as nylon or other man-made fibers and materials to the skin, or can cause a beta particle radiation burn to the skin (Cooray et al. 2015). Regardless, the most common skin burns in developed countries are quite superficial, at least to a medical person's view, and seldom need more than routine home care.

The lack of burns should not be used as evidence to eliminate lightning injury, which has occurred too many times in workers' compensation cases. Not all lightning victims have burns or marks of any kind. Less than half of lightning survivors in the Lightning Strike and Electric Shock Survivors International support group report any kind of skin marks at all. It is reasonable to expect that much of lightning energy has been dissipated before it reaches a person, especially for mechanisms such as ground current and contact injury. Unfortunately, it takes only a minuscule charge delivered at the right time in the cardiac cycle to cause death.

Occasionally, even in developed countries, lightning may cause burns significant enough to require grafting. In very rare cases, it may also cause actual tissue damage to internal organs such as the heart, as opposed to the more frequent simple interruption of the heart's normal electrical patterns and controls.

2.3 Lightning Injury in Developing Countries

In developing countries, where the vast majority of the housing is inadequate to provide safety, entire families, classrooms, or gatherings of people for church, work, and other activities can be injured or killed by lightning. This situation differs from that in the United States and other developed countries where the vast majority of injuries are to one or sometimes two people at a time (Curran et al. 2000; Cooper 2012; Holle 2016; Holle and Cooper 2016).

An additional and very significant risk is the common construction of most buildings in developing countries. Most buildings in developing countries, particularly in non-urban areas, are a combination of mud brick, cardboard, concrete block, or similar materials for the walls with roofs of thatch or sheet metal, sometimes held down by rocks. Homes are most often one or two room enclosures with no plumbing, electrical wiring, or metal studs in the walls that would provide lightning electricity a path to ground and harmlessly around the inhabitants, as housing in developed countries usually provides. In addition, the floor is usually packed earth instead of a material that would provide some insulation. Sometimes classrooms of students and church attendees meet outside or in an open area with only a metal roof above their heads. Agricultural workers, those at market or walking to market, school, and work, miners, and others may have no shelter at all; let alone "lightning safe" areas for escape (Holle 2016; Holle and Cooper 2016).

When lightning strikes in these situations, multiple injuries and deaths can occur. These are likely more newsworthy and more frequently reported than single incidents or those with small numbers of injuries, perhaps biasing the average number of people who are injured. However, regardless of any news reporting bias, the reports of ten-plus people injured in a single incident are quite different than the singletons usually reported in developed countries, making this a situation peculiar to developing countries and worthy of study and injury prevention messages that are tailored to these situations.

A particular risk in these types of structures is that the tinder-dry and generations-old thatch may catch fire. An effect of acute lightning injury, unknown to most people, is paralysis in up to two-thirds of victims that can last for minutes to hours, most often of the lower extremities (Cooper 1980). This paralysis can effectively prevent the escape of those inside as they watch the flaming thatch fall on them. The fire blocks rescue and those that might easily have survived, albeit with disability or other aftereffects, are reduced to charred bodies (Table 2.1). It is these cases that are quoted by reporters that may lead to the general misperception that all those killed by lightning are "burned beyond recognition."

Table 2.1 Newspaper articles of people burned inside African homes and a Colombian ritual building

http://www.timeslive.co.za/local/2017/01/25/%E2%80%98I-shouted-for-her-to-come-out-of-the-house%E2%80%99-Zumas-niece-survives-Nkandla-lightning-strike
http://www.monitor.co.ug/News/National/Lightning-kills-Amuru-couple/-/688334/2502700/-/5xmec9/-/index.html
http://www.cnn.com/2014/10/06/world/americas/colombia-lightning-strike/index.html

References

Anderson RB (2001) Does a fifth mechanism exist to explain lightning injuries? IEEE Eng Med Biol Mag 20(1):105–113

Anderson RB, Carte AE (2009) Struck by lightning. Archimedes 09:25–29

Anderson RB, Jandrell I, Nematswerani H (2002) The upward streamer mechanism versus step potentials as a cause of injuries from close lightning discharges. Trans SA Inst Elec Eng 93(1):33–43

Andrews CJ (1992) Telephone related lightning injury. Med J Aust 157(11/12):823–825

Andrews CJ, Darveniza M (1989) Telephone mediated lightning injury: an Australian survey. J Trauma 29(5):665–671

Blumenthal R (2012) Secondary missile injury from lightning strike. Am J Forensic Med Pathol 33:83–85

Blumenthal R, West NJ (2015) Investigating the risk of lightning's pressure blast wave. S Afr J Sci 111(3/4):2014–0187

Cooper MA (1980) Lightning injuries: prognostic signs for death. Ann Emerg Med 9(3):134–138

Cooper MA (2002) A fifth mechanism of lightning injury. Acad Emerg Med 9(2):172–174

Cooper MA (2012) Whether the medical aspects of lightning injury are different in developing countries. In: Preprints of the 4th international conference on lightning protection, Vienna, Austria, 2–7 Sept 2012

Cooper MA, Holle RL (2010) Mechanisms of lightning injury should affect lightning safety messages. In: Preprints of the 3rd international lightning meteorology conference, Vaisala, Orlando, Florida, 21–22 Apr 2010

Cooper MA, Holle RL, Andrews CJ (2008) Distribution of lightning injury mechanisms. In: Preprints of the 2nd international lightning meteorology conference, Vaisala, Tucson, Arizona, 21–23 April 2008

Cooper MA, Andrews CJ, Holle RL et al (2017) Lightning-related injuries and safety. In: Auerbach (ed) Wilderness medicine, 7th edn. Elsevier, Philadelphia, pp 71–117

Cooray V, Cooray GK, Cooray C (2015) On the possible mechanism of keraunographic markings on lightning victims. J Atmos Solar-Terrestrial Phys 136:119–123

Cummins KL, Krider EP, Olbinski M et al (2018) A case study of lightning attachment to flat ground showing multiple unconnected upward leaders. Atmos Res 202:159–174

Curran EB, Holle RL, López RE (2000) Lightning casualties and damages in the United States from 1959 to 1994. J Clim 13:3448–3453

Holle RL (2016) Lightning-caused deaths and injuries related to agriculture. In: Preprints of the 33rd international conference on lightning protection, Estoril, Portugal, 25–30 Sept 2016

Holle RL, Cooper MA (2016) Lightning-caused deaths and injuries at schools. In: Preprints of the 33rd international conference on lightning protection, Estoril, Portugal, 25–30 Sept 2016

Navarrete-Aldana N (2016) Lightning injuries in Colombia. From lack of awareness to prevention. Paper presented at the world meeting on lightning, Cartagena de Indias, Colombia, 6–8 Apr 2016

Rakov VA (2016) Fundamentals of lightning. Cambridge University Press, Cambridge

Saba MMF, Pinto O Jr, Ballarotti MG (2006) Relation between lightning return stroke peak current and following continuing current. Geophys Res Lett 33:L23807. https://doi.org/10.1029/2006GL027455

Chapter 3
Lightning Effects on the Body

Abstract Lightning can cause a large range of injuries. This chapter discusses diagnosis, reported injuries, listed by organ system, and whether they are common or uncommon injuries, as well as a brief discussion of treatment.

The most deadly injury is cardiac arrest at the time of the strike. In survivors, the most common immediate findings are amnesia, loss of consciousness, tympanic membrane rupture, keraunoparalysis, and other neurologic or musculoskeletal injuries. Burns are usually superficial in developed countries, although secondary burns from burning thatch may be fatal in developing countries where the majority of housing is unsafe.

The most common long term and disabling injuries are neurologic in nature: brain injury similar to post-concussive syndrome, and chronic pain. For a greatly expanded discussion of lightning injuries, including treatment, and for much more extensive references than we can include here, see Cooper et al. (Lightning-related injuries and safety. In: Auerbach PS (ed) Wilderness medicine, 7th. Elsevier, Philadelphia, pp 71–117, (2017)).

3.1 Diagnosis

Diagnosis of lightning injury is easy when there are witnesses and thunderstorms are in the area. However, it may not be the first diagnosis to come to mind if there were no witnesses or no recognition of thunderstorms in the area. Clues to the diagnosis are linear burns, mental status changes, ruptured tympanic membranes ("eardrums"), clothes exploded off, and the relatively rare Lichtenberg figures (Resnik 1996; Cherington et al. 2003). While the vast majority of lightning injuries in the developed world are to people who are outdoors, this may not be true of developing countries for reasons discussed in several other chapters.

Although mechanisms of injury by lightning are discussed in Chap. 2, it is seldom useful for the physician to expend time trying to determine the mechanism or "where the lightning traveled," either in the environment or in the person. Unlike other trauma, where knowing details about how the injury occurred can be quite helpful, there is no evidence that knowing the mechanism aids in diagnosis and care

M. A. Cooper, R. L. Holle, *Reducing Lightning Injuries Worldwide*,
Springer Natural Hazards, https://doi.org/10.1007/978-3-319-77563-0_3

of lightning victims. In reality, there is little hope of validating hypotheses about pathway and mechanism at the time of treatment. Examining lightning data and the scene after the survivors have been treated may identify where lightning protection may potentially be useful or where it has failed and where lightning injury prevention education may be needed. However, the physician's prime concern at the time of injury should always be the care of the victims, not scene or mechanism reconstruction.

Errors in diagnosis and treatment occur when physicians expect to find injuries based on extrapolation from their knowledge of household current or high-voltage injury. Unfortunately, physicians seldom have a good understanding of the physics of electricity or of electric fields, particularly the peculiar physics and characteristics of lightning. Unless they are in the lightning protection field, electrical engineers are unlikely to have knowledge or experience with the peculiarities of lightning or familiarity with international lightning protection codes so that their opinions should not be relied on for forensic investigations.

A Makeshift Visual Analogy
Lightning injury is more like being sprayed with a huge hose (multiple areas of simultaneous impact) than it is like being shot by a gun or arrow. A path of injury, such as occurs with an arrow, is a not useful concept for lightning. There are no "entry" and "exit" points with lightning as there are with gunshot wounds and high-voltage injuries. Besides, with rapidly reversing (alternating) current electrical injuries, it would make much more sense to call these "source" and "ground" than exit/entry.

3.2 Early Studies

While there is a long series of simple case reports and anecdotal papers about lightning injury, prior to Cooper's study in 1980, there were no studies on prognosis or on long-term injuries (Cooper 1980). This paper was the first substantial case series or meta-analysis of lightning injuries and helped define lightning injury for the next decade or more (Table 3.1). Andrews et al. (1989) analyzed a similar case mix; while the numbers in this study were a bit different for some of the signs, overall findings corroborated the 1980 study. It should be noted that studies based on case reports are likely to overestimate the incidence of signs and symptoms as only cases with injuries will be published, and survivors with few or no findings will be excluded from the publications.

The case reports were coded for the many reported signs and symptoms. How to tell "where the lightning had traveled" in or on the body was unknown, so the locations of the reported burns were coded. This simple analysis showed significant correlations between:

Table 3.1 Results of the first organized analysis of lightning injury in the medical literature

Finding	Number affected/cases reporting sign or symptom	%
Death	20/66	30
Survival with sequelae	20/27	70
Cardiopulmonary arrest	17/58	30
Loss of consciousness	54/61	72
Confusion/amnesia	24/28	86
Paralysis	20/29	69
Leg paralysis	20/29	69
Arm paralysis	11/29	30
Burns		
Head	24/54	44
Trunk	36/54	66
Arm	16/54	30
Leg	30/54	55
Multiple burn locations	34/54	63
Single burn locations	14/54	26
No burns	6/54	11
Not reported	12/66	18
Pregnancy	9 cases reported	
Maternal death	9	
Fetal death in utero	2	
Neonatal death	2	
Apparently healthy newborn	4	
Unreported fetal outcome	1	

Cooper 1980

1. Leg burns and death – Victims with leg burns were five times more likely to die than those without ($p < 0.5$). There were no significant correlations between arm-to-arm, arm-to-leg, or head-to-leg burns.
2. Cranial burns and death – Victims with burns to the head were three to four times more likely to die ($p < 0.25$).
3. Cranial burn and cardiac arrest – Victims with burns about the head were more likely to have a cardiac arrest ($p < 0.25$).
4. Cardiopulmonary arrest and death – The *only* victims in the analysis who died were those who suffered cardiopulmonary arrest at the time of the injury ($p < 0.0001$).

Medical Definitions

Complaints: Nonpejorative medical term for what patients tell a medical worker they have ("complains of") as symptoms. It does not mean that the patient is complaining or whining.

Sequela (plural sequelae): A pathological condition resulting from a disease, injury, therapy, or other trauma. Examples are paralysis after a stroke, limp after a broken leg, chronic pain after nerve injury, and scars after a serious burn.

These correlations have withstood over 35 years of scrutiny and remain valid today, although it took a decade or more to appreciate that leg burns might be from ground current.

3.3 Risk of Lightning Injury

Lightning injury is not simply an unpredictable act of nature. There are specific factors that are useful in predicting the risk of lightning injury. Table 3.2 shows a quick synopsis of the major factors that determine the risk of being injured by lightning (Chap. 6).

3.4 Injuries from Lightning

In the developed world, the proximate and only cause of death from lightning is cardiac arrest at the time of the injury (Cooper 1980). The legal pronouncement of death may be delayed a few days if the patient is resuscitated but has irreparable brain damage. Death from suicide may occur in some survivors months to years

Table 3.2 Factors that determine the risk of being injured by lightning

Exposure
Lightning density, measured in the number of cloud-to-ground flashes/square km/year
Population density
Availability of safe areas to escape lightning
Existence of lightning safety guidelines
Knowledge of lightning safety guidelines
Compliance with guidelines

Table 3.3 Common immediate and delayed signs and symptoms from lightning injury

Immediate signs and symptoms	Delayed signs and symptoms
Cardiac arrest and cardiac injuries	Tinnitus, hearing loss
Pulmonary injuries	Balance problems
Neurologic signs, seizures	Neurologic symptoms and signs
Deafness, usually temporary	Neuropsychological changes
Confusion, amnesia	Memory coding, processing, and accessing
Blindness – often temporary	Attention deficit
Dizziness and balance problems	Loss of executive function
Organ contusions from shock wave	Distractibility
Secondary "shrapnel" wounds	Personality changes
Chest pain, muscle aches	Irritability
Tympanic membrane rupture	Chronic pain syndromes
Headache, nausea	Musculoskeletal pain
Post-concussion syndrome	Changes in vision
Photophobia and hyperacusis	Seizures

after the injury because of despair from loss of abilities, loss of work, social and family relationships, chronic pain, and other factors. When adequate medical care is unavailable, survivors may self-medicate with alcohol or other drugs.

In the developing world, secondary causes of death may be from fiery thatch falling on the family if lightning paralysis (keraunoparalysis) or other factors keep them from evacuating their home or workplace.

Table 3.3 shows immediate versus long-term problems. Selected references for injuries for each organ system are listed at the end of this chapter. Many more are available in the medical literature but are often isolated case reports or small case series, not research studies that are useful for treatment and long-term recovery.

3.4.1 *Cardiovascular System*

Cardiovascular systems signs are identified in Table 3.4.

3.4.2 *Lung-Pulmonary Injuries*

Acute pulmonary injuries have been reported but are relatively uncommon except as complications of cardiac arrest and resuscitation (Table 3.5).

Table 3.4 Cardiac signs with lightning injury

	Immediate common	Immediate uncommon	Delayed uncommon
Cardiac arrest		**X(10%?)**	
Atrial fibrillation		**X**	
Asystole (standstill)		**X**	
Other arrhythmias		**X**	
ECG abnormalities		**X**	X
Prolonged QT interval		**X**	
Coronary artery spasm		**X**	
Myocardial ischemia	**X**		
Focal necrosis		**X**	
Cardiogenic shock		**X**	
Cardiomyopathy		**X**	X
Aneurysmal dilatation			X
Hypertension	**X**		X

Table 3.5 Pulmonary injuries

	Immediate uncommon	Immediate very rare
Blunt injury	**X**	
Contusion	**X**	
Hemoptysis		**X**
Pneumomediastinum		**X**
Pulmonary edema	**X**	
Pulmonary hemorrhage		**X**

3.4.3 Nervous System Injury

As noted, the vast majority of injury from lightning is to the nervous system, both acutely and in the long term. These can be debilitating injuries, destroying a person's ability to return to their prior work, and putting stress on families both emotionally and economically. Please note the added column on the right of Table 3.6 to denote that many of these injuries are permanent.

3.4.4 Neuropsychological Injuries from Lightning

When the term "psychological injury" is used, many people think that it means only a mood disorder such as depression or unusual behavior. Many also think these problems are in the victim's control. However, mood disorders are only the "tip of

Table 3.6 Nervous system injuries from lightning

	Immediate common	Immediate uncommon	Delayed common	Delayed rare	Prolonged or permanent
Post-traumatic headache	**X**				X
Skull fracture		**X**			
Sleep disturbances	**X**				X
Brain					
Amnesia	**X**				X
Anoxic brain injury		**X**			X
Aphasia		**X**			X
Ataxia		**X**			X
Brain injury similar to post-concussive syndrome	**X**				X
Cerebellar ataxia, acute or delayed		**X**		X	
Cerebellar infarction		**X**			
Confusion	**X**				X
Incoordination	**X**				
Intracranial hemorrhage		**X**			
Loss of consciousness	**X**				X
Movement disorders and Parkinsonism				X	X
Paresis	**X**				X
Seizures		**X**			
Syndrome of Inappropriate antidi- uretic hormone		**X**			
Cranial nerves					
Facial nerve palsies		**X**			
Tinnitus	**X**				X
Dizziness, balance problems	**X**				X
Spinal cord					
Hemiplegia		**X**			X
Keraunoparalysis	**X**				
Progressive muscle atrophy of the upper extremities				X	
Quadriplegia, acute or delayed		**X**		X	X
Spinal artery spasm	**X**				
Autonomic nervous system injury					
Complex regional pain syndrome				X	X
Diffuse degeneration				X	X

(continued)

Table 3.6 (continued)

	Immediate common	Immediate uncommon	Delayed common	Delayed rare	Prolonged or permanent
Hypertension	**X**			X	X
Cardiac arrhythmias		**X**			X
Peripheral nervous system					
Pain	**X**		X		X
Paresthesia	**X**		X		X
Endocrine dysfunction					
Amenorrhea / men- strual irregularities				X	
Central hypoadrenergic state				X	X
Decreased libido			X		X
Hypoadrenalism				X	X
Hypogonadism				X	X
Hypopituitarism				X	X

the iceberg," and behaviors are only an observable outward sign of unrest or damage within the person.

When brain injury occurs, whether from blunt injury from a fall, from whipping the head around in a car accident, from sports injury, or from lightning, there may be physically observable deficits such as hematomas and contusions, "diffuse axonal injury" with micro-tears to the brain substance, or subvisible (by current imaging techniques) damage. All of these can result in deficits in the normal function of the brain, including all types of memory processing deficits such as coding of new information, accessing stored information, "working memory" used to follow directions, holding more than one memory at the same time which will affect multitasking, organization and executive function, etc. Table 3.7 lists deficits reported in the literature, by patients, and corroborated on neuropsychological testing, a 4–8 h battery of pen-and-paper or computer-aided testing of IQ, memory, processing speed, and many other factors administered by a trained technician and interpreted by a board-certified neuropsychologist. Neuropsychological testing is expensive and not always useful unless needed for secondary reasons such as cognitive rehabilitation or legal cases.

Neuropsychological corroboration is listed as "delayed" because, while these problems may be appreciated by the survivor within a few weeks or months of the injury, neuropsychological testing, if done at all, is often delayed for months to years. Almost all of these deficits are permanent and may affect the survivor's ability to return to work, their personal relationships, and their self-worth (Cooper 2001; Cherington 2003, 2005a, b; Andrews 2006). Recently, DSM (*Diagnostic and Statistical Manual of Mental Disorders*) criteria have been proposed that may aid in recognition, diagnosis, and corroboration of lightning injury (Andrews 2017).

Table 3.7 Psychological and thought-processing disorders from lightning injury

	Immediate common	Delayed common
Reported in studies		
Photophobia/photosensitivity	**X**	
Hyperacusis/sensitivity to sound	**X**	
Agoraphobia/avoidance of crowds	**X**	X
Emotional lability		X
Mood abnormalities		X
Post-traumatic stress disorder		X
Sleep disturbance	**X**	X
Anxiety, hyper-vigilance	**X**	X
Executive function loss	**X**	X
Deficits in "working memory"	**X**	X
Self-reported symptoms		
General memory problems		X
Concentration deficit	**X**	X
Loss of "mental power"	**X**	X
Aggression/personality change/ irritability		X
Storm phobia	**X**	X
Low libido		X
Social isolation		X
Vocabulary and difficulty finding word		X
Low mood/depression		X
Learning dysfunction		X
Anxiety	**X**	X
Marital stress		X
Deficits validated by neuropsychological testing		
Auditory memory		X
Processing speed		X
Vocabulary/word finding/verbal learning		X
Verbal fluency deficit		X
Visual memory deficit		X
Concentration loss		X
Executive and cognitive processing loss		X
Loss of attention span		X
Anxiety		X
General memory deficit		X
Visuospatial deficit		X
Verbal IQ loss		X
Decrease in IQ		X

Table 3.8 Eye injuries from lightning

	Immediate common	Immediate uncommon	Delayed common	Delayed uncommon
Anisocoria		**X**		
Blindness	**Transient**			
Cataracts	**X**	**X**	X	
Choroidal rupture		**X**		
Chorioretinitis		**X**		
Corneal lesions		**X**		X
Decreased color sense		**X**		
Diplopia		**X**		
Horner's syndrome		**X**		
Hyphema		**X**		
Iridocyclitis		**X**		
Iritis		**X**		
Loss of accommodation		**X**		
Loss of light reflexes		**X**		
Macular holes	X			
Macular degeneration		**X**		
Mydriasis	**X**			
Optic atrophy		**X**		
Optic neuritis		**X**		
Photophobia	**X**			
Retinal separation		**X**		
Uveitis	**X**			
Vitreous hemorrhage	**X**	**X**		

While any type of disability can be difficult on marital and other close relationships, neuropsychological disabilities are peculiar in that they often result in personality changes and hyperirritability, especially toward loved ones, which can lead to the breakup of the relationships closest to the survivor (Cooper 2001; Cooper and Marshburn 2005).

3.4.5 Lightning Injuries to the Eyes and Ears

Lightning injury often affects the vision and hearing as well as the structures near the eyes and ears as shown in Tables 3.8 and 3.9.

Table 3.9 Ear injuries from lightning

	Immediate common	Immediate uncommon	Length of symptoms
Blast injury	X		Resolves over time
Deafness	X		Usually temporary
Tinnitus	X		Usually permanent
Facial nerve palsies		X	
Hemotympanum (blood behind the eardrum)		X	Investigate underlying trauma
Occult fractures of the jaw or styloid process.		X	
Ossicular (ear bones) or mastoid disruption.		X	May be difficult to manage
Otorrhea (cerebrospinal fluid leak)		X	Difficult to manage
Sensorineural damage to eighth cranial nerve with hearing changes, tinnitus (ringing in the ears), ataxia, and dizziness/balance problems	X		Usually needs physical therapy; some will be permanent
Tympanic membrane (ear drum)	X		Often heals spontaneously rupture

3.4.6 *Lightning Injury to the Skin*

Burns in Developed Countries Burns and skin injury, when they occur, are usually superficial (leaving no scar) and relatively minor, at least to the physician's view. Survey of the Lightning Strike and Electric Shock Survivors International support group showed that only about one-third of them had skin marks of any kind from their injury. This is probably due to two factors:

1. Most people are not injured by a direct strike, sideflash, or upward streamer; the mechanisms most likely to have thermal injury. Ground current and contact injury are indirect mechanisms and may have an electrical effect, but not always enough thermal energy to cause a burn.
2. Flashover, where the majority of the energy flows rapidly around the person, not through them, and the rapidity of lightning (seldom lasting more than a few microseconds) means that lightning is simply not around long enough to cause significant heating or breakdown of the skin.

Often, the damage that is caused to the skin is "secondary" injury when sweat or rainwater on the skin turns to steam, because of the flashover mechanism, and the hot vapor causes burns. Sometimes, metal in the clothing is heated and causes secondary discharge to the skin underneath. Burns, when they occur, can be divided into five types: linear burns, punctate, full-thickness burns, Lichtenberg flowers, thermal burns from ignited clothing or heated metal, and combinations as shown in Figs. 3.1, 3.2, 3.3, 3.4, and 3.5.

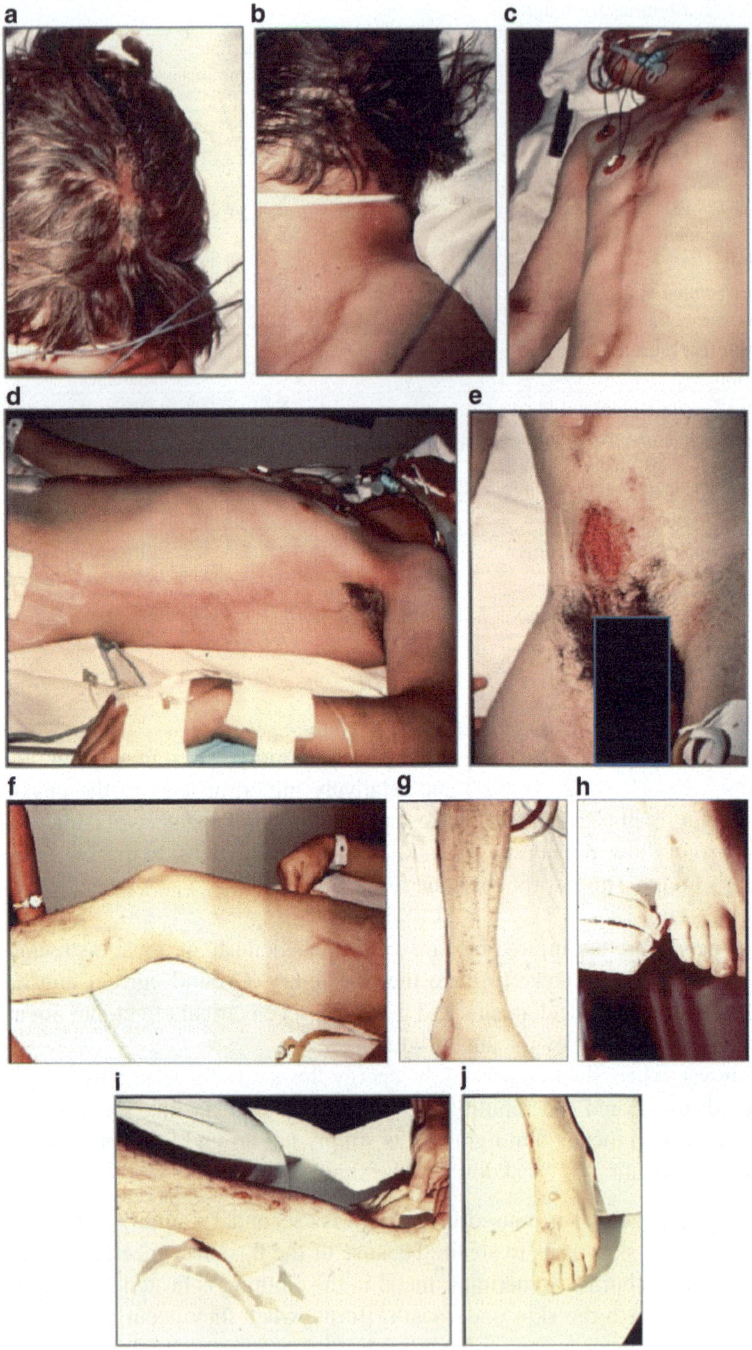

Fig. 3.1 Linear burns from fatal 1977 lightning injury to 22-year-old baseball player. Most of these marks did not appear until a few hours to a few days (those with small eschar – "scab") after the injury. (**a**). Mark that matured on the back of the head by the third day. (**b**) Linear marks

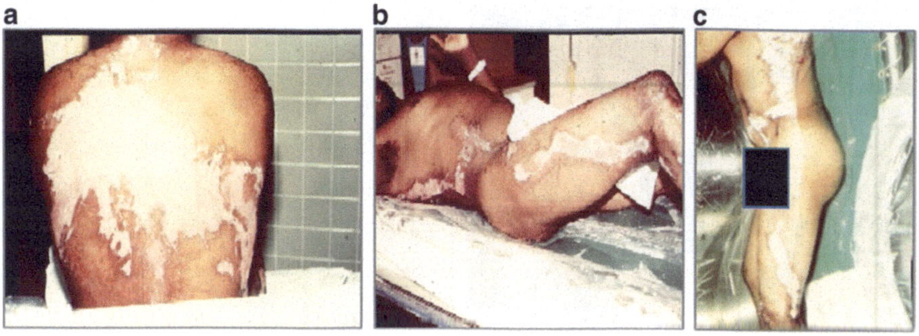

Fig. 3.2 Steam burns on a motorcyclist who was wearing a leather jacket, belt, and pants. (**a**) Note the more extensive burn to the back where the steam was held against the skin longer by the leather jacket than would occur with more porous clothing. (**b**, **c**) More linear burns where the jacket was not as close to the skin. Note that the burns take a right angle under the belt at the waist before splitting at the groin and continuing laterally down the leg. Old scarring on the right knee was from a previous airplane crash and not part of the lightning injury. (Photos ©M.A. Cooper)

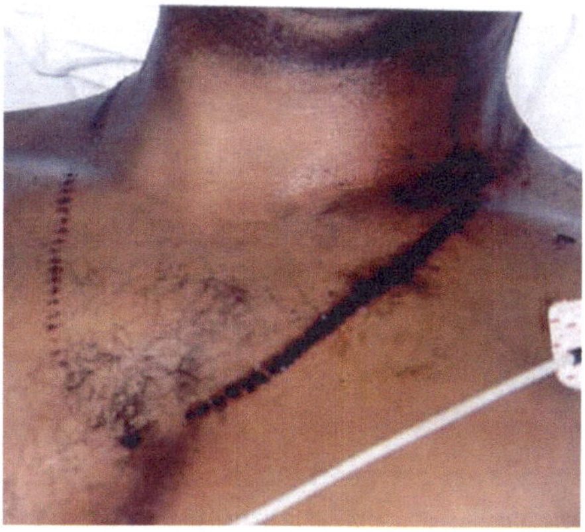

Fig. 3.3 Metal necklace burned into skin by lightning with permanent tattooing (nonfatal injury). (Courtesy R. Washington. (Photo ©M.A. Cooper))

Fig. 3.1 (continued) continuing down the side of the neck. (**c**, **d**) Continuing marks down the anterior and lateral torso – note that these are the normal "sweat lines" that a baseball player in July would develop. Also note the burn to the antecubital (elbow) fossa where sweat would accumulate under the baseball jersey as the player stood crouched and ready to catch near second base. Note that all of these burns are partial thickness with sparse blistering. (**e**) More extensive burns where a metal belt buckle or athletic supporter may have been causing secondary thermal burns or electrical discharge to the skin from the metal. (**f–j**). Damage to legs and feet. Note the parallel marks on the foreleg which would correspond to sweat accumulation or wetness in the ribbing of the athletic socks. Note also the mark on the heel which may have been from the metal heel cup in the shoe or contusion from the shoe being ripped off by the vapor explosion of the flashover. The socks were destroyed or exploded off below the ankle, and the shoes were never found. Blunt injury from the explosion caused non-burned ripping of the fifth toe web as well (37H). (Photos ©M.A. Cooper)

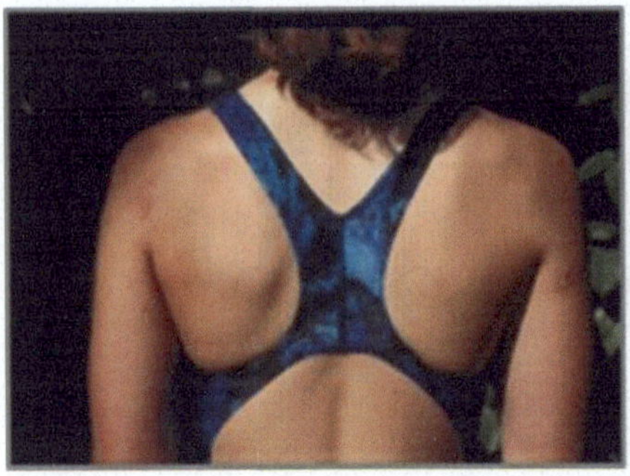

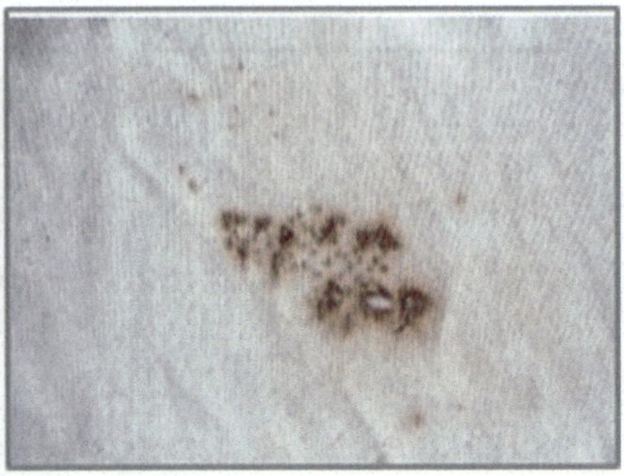

Fig. 3.4 (**a**) Punctate burns to the shoulder and arm of an emergency physician's 12-year-old daughter who was at a campfire with friends. (**b**) Note singing of the cotton T-shirt that she was wearing at the time over the area of the right arm burn. (Photos ©M.A. Cooper)

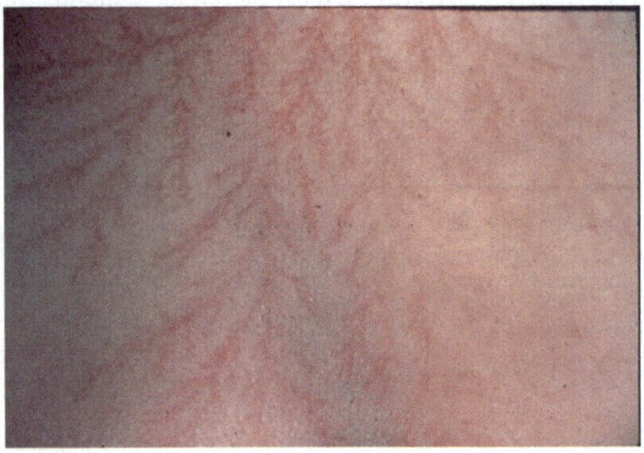

Fig. 3.5 Lichtenberg figure or feathering marks. (Photo ©M.A. Cooper)

A peculiar mark that is **not** a burn is the Lichtenberg figure (also called Lichtenberg flowers, arborescent marks, fractals, ferning) (Fig. 3.5). However, nothing except lightning causes this pattern, so it is diagnostic (pathognomonic) but usually lasts only a few hours to days before it fades.

Burns in Developing Countries In developing countries, reports of deaths may describe lightning victims as "charred" or "burned beyond recognition." Although these reports were formerly thought to be simply reports written by reporters with no first-hand knowledge who might have expected this level of injury to be standard, several newspaper reports from various parts of the world note gathered rescuers hearing screams from inhabitants but hampered from rescuing them due to the intensity of the tinder-dry, burning thatch (Table 2.1). The explanation may be that the keraunoparalysis that occurs acutely with lightning may prevent otherwise healthy individuals from being able to evacuate as the burning thatch falls on them.

3.5 Distribution of Injuries

While we know the range of injuries that survivors may suffer, one piece of data that is not collected anywhere is the distribution of injury severities of the survivors. Some will have insignificant shocks with little or no aftereffects. Others will have the typical brain injury and chronic pain syndromes that may render them unable to return to their prior line of work. Still others will have had cardiac arrest with brain injury from lack of oxygen and potentially be nearly vegetative for their rest of their lives. The distribution of survivor injuries is unknown and not collected by any means at this time.

While some survivors may need little aftercare, others will need significant rehabilitation or even require support for the rest of their lives. In developed countries, this is usually recognized, and care is available. However, in developing countries, little, if any, attention is given to survivors after the initial news report. Survivors' aftereffects may not be recognized, and there may be no care or rehabilitation afterward, leaving individuals suffering and families at a loss when their child has a learning disability after brain injury or the family provider becomes unable to return to work (Cooper 2001).

3.6 Acute Treatment of Lightning Injuries

Rescuers should immediately call for help and activate any emergency response such as 911, if it is available, so that they have more help if several people are injured. Cardiac arrest at the time of the injury is the **only** cause of death, so, when the number of victims overwhelms the number of rescuers, the "rule" is to take care of those most severely injured because anyone moaning, groaning, or breathing will survive even if they have permanent or disabling aftereffects.

With the exception of the indoor burn injuries in developing countries, the vast majority of those who are injured by lightning are injured outdoors. If thunderstorms are in the area, rescuers will be at risk. Some victims may need cardiopulmonary resuscitation (CPR); others will be responsive, even walking around or sitting dazed. CPR should be started for unresponsive victims who have little or no breathing. Automatic external defibrillators (AEDs), if available, have been helpful in some cases, but resuscitation should not be delayed to find one (Cooper and Johnson 2005; Nelson et al. 2007; Vanden Hoek et al. 2010).

The probability that lightning victims will recover after prolonged cardiopulmonary resuscitation is low, so if there is no response after 20–30 min of CPR, it is reasonable to halt resuscitation efforts, as it is unlikely that further efforts will be effective. Other routine first aid, sheltering the person from rain that can cause hypothermia, and stabilization of the scene for the protection of all involved should be done.

3.7 Emergency Department Evaluation and Treatment

Emergency department treatment is routine: resuscitate if in arrest, stabilize if unstable, evaluate and treat general and individual injuries, decide if any of the injuries or level of consciousness indicate admission, and call in specialists, if needed. The vast majority of those struck by lightning will not need to be admitted and can safely be discharged to the care of a responsible relative. Unfortunately, we know very little about interventions that will change the development of the aftereffects of lightning injury. For a more extensive discussion of assessment and treatment, see Cooper et al. (2017).

If there are breaks in the skin or burns, tetanus immunization should be updated. Antibiotics are not necessary until there are specific signs that infection is developing days later.

There is no specific treatment for lightning that we know. Care is standard for the cardiac, pulmonary, and other injuries that may occur. If lightning did not cause immediate cardiac arrest, there is very low risk of death. However, observation, cardiac monitoring, and serial cardiac marker measurements are indicated if there is any sign of cardiac ischemia or arrhythmia or if the victim complains of chest pain. No systematic study on the duration of observation has been undertaken for lightning survivors. Transient hypertension from autonomic instability and injury may be so short-lived as to not require acute therapy.

3.8 Long-Term Care

In the long run, there is no specific treatment for lightning injury. Lightning is a nervous system injury that can involve chronic pain, neuropathy, and brain injury, sometimes complicated by initially unrecognized musculoskeletal injury. Most symptoms can be treated in standard fashion, including cognitive therapy, pain management, job retraining, and counseling as indicated by the survivor's signs and symptoms. As with any serious illness, the caregiver and family will be the unsung heroes and need support, recognition, and counseling, as well as respite.

3.9 Referral to Support Groups and Other Information Sources

Lightning Strike and Electric Shock Survivors International (LSESSI), a support group founded in the late 1980s, has numerous materials for survivors and their families. LSESSI is located at P.O. Box 1156, Jacksonville, NC 28541-1156, telephone 910-346-4708, website http://www.lightning-strike.org) (Cooper and Marshburn 2005). Another useful website is http://www.lightningsafety.noaa.gov.

References and Reading Lists by Organ System

General

Andrews CJ, Darveniza M, Mackerras D (1989) Lightning injury: a review of clinical aspects, pathophysiology and treatment. Adv Trauma 4:241–287

Andrews CJ, Cooper MA, Darveniza M et al (1992) Lightning injury: electrical, medical, and legal aspects. CRC Press, Boca Raton

Cooper MA (1980) Lightning injuries: prognostic signs for death. Ann Emerg Med 9(3):134–138

Cooper MA (2016) Lightning injuries. eMedicine: An on-line emergency medicine text 2011. Available via Medscape. https://emedicine.medscape.com/article/770642-overview. Accessed 22 Oct 2017

Cooper MA, Andrews CJ, Holle RL et al (2017) Lightning-related injuries and safety. In: Auerbach PS (ed) Wilderness medicine, 7th edn. Elsevier, Philadelphia, pp 71–117

Cardiac Injuries

Alyan O, Ozdemir O, Tufekcioglu O et al (2006) Myocardial injury due to lightning strike – a case report. Angiology 57(2):219–223

Andrews CJ, Colquhoun D (1993) The QT interval in lightning injury with implications for the 'cessation of metabolism' hypothesis. J Wilderness Med 4:155–166

Andrews CJ, Darveniza M (1989) Effects of lightning on mammalian tissue. In: Preprints of the international conference on lightning and static electricity, Bath, United Kingdom, 26–28 Sept 1989

Auer J (2000) Images in cardiology. Cardiac involvement in lightning strike injury. Clin Cardiol 23(5):386

Cooper MA, Johnson SA (2005) Cardiopulmonary resuscitation and early management of the lightning strike victim. In: Ornato JP, Peberdy MA (eds) Cardiopulmonary Resuscitation. Humana Press, Totowa

Cooper MA, Holle R, Andrews C (2008) Electrical current and lightning injury. In: Field J (ed) The textbook of emergency cardiovascular care and CPR: ACLS for the experienced provider. Williams & Wilkins, New York

Cooray V, Cooray C, Andrews CJ (2007) Lightning caused injuries in humans. J Electrost 65(5–6):386–394

Dronacahrya L, Poudel R (2008) Lightning induced atrial fibrillation. Kathmandu Univ Med J 6(24):514–515

Emet M, Aksakal E, Aslan S et al (2010a) Lightning-induced lethal myocardial infarction and ventricular tachycardia in a boy. Anadolu Kardiyol Derg 10(4):378–379

Grubb BP, Karabin B (2007) New onset postural tachycardia syndrome following lightning injury. Pacing Clin Electrophysiol 30(8):1036–1038

Hayashi M, Yamada H, Agatsuma T et al (2005) A case of Takotsubo-shaped hypokinesis of the left ventricle caused by a lightning strike. Int Heart J 46(5):933–938

Jackson SH, Parry DJ (1980) Lightning and the heart. Br Heart J 43(4):454–457

Karadas S, Vuruskan E, Dursan R et al (2013) Myocardial infarction due to lightning strike. J Pak Med Assoc 63(9):1186–1188

Lichtenberg R, Dries D, Ward K et al (1993) Cardiovascular effects of lightning strikes. J Am Coll Cardiol 21(2):531–536

Leiria T, Pires L, Kruse M et al (2013) Struck by lightning: a case of nature-induced pre-excited atrial fibrillation. Circ Arrhythm Electrophysiol 6(2):e20–e21

McIntyre WF, Simpson CS, Redfearn DP et al (2010) The lightning heart: a case report and brief review of the cardiovascular complications of lightning injury. Indian Pacing Electrophysiol J 10(9):429–434

Nelson KL, Mills W Jr, Umbel S et al (2007) Lightning, sudden cardiac death, simulation and an automated external defibrillator: the perfect storm. Resuscitation 74(3):567–571

Palmer AB (1987) Lightning injury causing prolongation of the Q-T interval. Postgrad Med J 63(744):891–894

Rivera J, Romero KA, Gonzalez-Chon O et al (2007) Severe stunned myocardium after lightning strike. Critical Care Med 35(1):280–2285

Slesinger T, Bank M, Drumheller B et al (2010) Immediate cardiac arrest and subsequent development of cardiogenic shock caused by lightning strike. J Trauma 68(1):E5–E7

Vanden Hoek TL, Morrison L, Shuster M et al (2010) American heart association guidelines for cardiopulmonary resuscitation and emergency cardiovascular care, part 12: Cardiac arrest in special situations. Circulation 122(18 Supp 3):S829–S861

Zack F, Hammer U, Klett I et al (1997) Myocardial injury due to lightning. Int J Legal Med 110(6):326–328

Pulmonary Injuries

Bouwen L, Bosmans E (1997) Posttraumatic pneumomediastinum: not always a cause for alarm. Acta Chir Belg 97:145–147

Halldorsson A, Couch MH (2004) Pneumomediastinum caused by a lightning strike. J Trauma 57(1):196–197

Moulson AM (1984) Blast injury of the lungs due to lightning. Br Med J (Clin Res Ed) 289(6454):1270–1271

Soltermann B, Frutiger A, Kuhn M (1991) Lightning injury with lung bleeding in a tracheotomized patient. Chest 99(1):240–242

van Waes OJF, van de Woestijne PC, Halm JA (2014) "Thunderstruck": penetrating thoracic injury from lightning strike. Ann Emerg Med 63(4):457–459

Nervous System Injuries

Aggaswal A, Pravin S (2011) Bolt from the blue: basal ganglion bleed following lightning strike. Surg Neurol Int 2:170

Cherington M (1995) Central nervous system complications of lightning and electrical injuries. Semin Neurol 15(3):233–240

Cherington M (2003) Neurologic manifestations of lightning strikes. Neurology 60(2):182–185

Cherington M (2005a) Neurorehabilitation of the multifaceted and complicated neurologic problems associated with lightning and electrical injuries. NeuroRehabilitation 20(1):1–2

Cherington M (2005b) Spectrum of neurologic complications of lightning injuries. NeuroRehabilitation 20(1):3–8

Cherington M, Yarnell PR (2003) Ball lightning encephalopathy. J Burn Care Rehabil 24(3):175

Cherington M, Yarnell P, Hallmark D (1993) MRI in lightning encephalopathy. Neurology 43(7):1437–1438

Cherington M, Yarnell P, Lammereste D (1992) Lightning strikes: nature of neurological damage in patients evaluated in hospital emergency departments. Ann Emerg Med 21(5):575–578

Cherington M, Walker J, Boyson M et al (1999) Closing the gap on the actual numbers of lightning casualties and deaths. In: Preprints of the 11th conference on applied climatology, American Meteorological Society, Dallas, Texas, 10–15 Jan 1999
Cohen JA (1995) Autonomic nervous system disorders and reflex sympathetic dystrophy in lightning and electrical injuries. Semin Neurol 15(4):387–390
Desai B, Fairclough R (2011) A case of a speech impediment following a near lightning strike. Int J Em Med 4:60
Emet M, Caner I, Cakir M et al (2010b) Lightning injury may cause abrupt cerebral salt wasting syndrome. Am Emerg Med 28(5):640.e1–640.e3
Fox M, Gubler C (2009) Struck dysphagic. Gastroenterology 137(3):E9–10
Hawkes C, Thorpe J (1992) Acute polyneuropathy due to lightning injury. J Neurol Neurosurg Psychiatry 55(5):388–390
Hendler N (1994) Reflex sympathetic dystrophy and causalgia. In: Tollison CD (ed) Handbook of pain management. Will & Wilk, Baltimore
Hendler N (2005) Overlooked diagnoses in chronic pain: analysis of survivors of electric shock and lightning strike. J Occup Environ Med 47(8):796–805
Hendler N, Kozikowski J (1993) Overlooked physical diagnoses in chronic pain patients involved in litigation. Psychosomatics 34(6):494–500
Hendler N, Bergson C, Morrison C (1996) Overlooked physical diagnoses in chronic pain patients involved in litigation, part 2: the addition of MRI, nerve blocks, 3d ct, and qualitative flow meter. Psychomatics 37(6):509–517
Jost WH, Schonrock LM, Cherington M (2005) Autonomic nervous system dysfunction in lightning and electrical injuries. NeuroRehabilitation 20(1):19–23
Kint PA, Stroy JP, Parizel PM (1999) Basal ganglia hemorrhage secondary to lightning stroke. JBR-BTR 82(3):113
Kleiter I, Luerding R, Diendorfer G et al (2007a) A lightning strike to the head causing a visual cortex defect with simple and complex visual hallucinations. J Neurol Neurosurg Psychiatry 8(4):423–426
Lakshminarayanan S, Chokroverty S, Eshkar N et al (2008) The spinal cord in lightning injury: a report of two cases. J Neurol Sci 276:199
Lammertse DP (2005) Neurorehabilitation of spinal cord injuries following lightning and electrical trauma. NeuroRehabilitation 20(1):9–14
O'Brien CF (1995) Involuntary movement disorders following lightning and electrical injuries. Semin Neurol 15(3):263–267
Ohashi M, Hosoda Y, Fujishiro Y et al (2001) Lightning injury as a blast injury of skull, brain, and visceral lesions: clinical and experimental evidences. Keio J Med 50(4):257–262
Parsaik AK, Ahlskog JE, Singe RW et al (2013) Central hyperadrenergic state after lightning strike. Clin Auton Res 23(4):169–173
Piccolo AP, Pedroso DFF, Moraes V et al (2007) Rhabdomyolysis and nerve deficits after lightning injuries—Review of the literature and a case report. Burns 33(1, Supplement):S25
Raja SN (1990) Sympathetically maintained pain. Curr Prac Anaesth, Mosby 2:421–425
Ramati A, Pliskin NH, Keedy S et al (2009) Alteration in functional brain systems after electrical injury. J Neurotrauma 26(10):1815–1822
Reisner AD (2013) Possible mechanisms for delayed neurological damage in lightning and electrical injury. Brain Inj 27(5):565–569
Schwartzman RJ (1994) Causalgia and reflex sympathetic dystrophy. In: Feldmann E (ed) Current diagnosis in neurology. Mosby, St Louis
Schwartzman R, Kerrigan J (1990) The movement dosorder of reflex sympathetic dystrophy. Neurology 40:57–61
Stanley LD, Suss RA (1985) Intracerebral haematoma secondary to lightning stroke: case report and review of literature. Neurosurgery 16(5):686
Yarnell PR (2005) Neurorehabilitation of cerebral disorders following lightning and electrical trauma. NeuroRehabilitation 20(1):15–18

Psychologic Disorders

Andrews CJ (2003) SPECT scanning and neuropsychological testing in electrical injuries. In: Proceedings of the international conference on lightning and static electricity, Royal Aeronautical Society, Blackpool, United Kingdom, 16–18 Sept 2003

Andrews CJ (2006) Further documentation of remote effects of electrical injuries, with comments on the place of neuropsychological testing and functional scanning. IEEE Trans Biomed Eng 53(10):2102–2113

Andrews CJ (2012) The origin of remote symptoms in electrical and lightning injuries: an attempt at explanation and a hypothesis for testing. J Lightnin Res (online) 1(1):1

Andrews CJ (2014) A review of the psychological consequences of lightning and electrical injury. In: Proceedings of the 32nd international conference on lightning protection, Shanghai, China, 11–18 Oct 2014

Andrews CJ, Reisner A (2017) Neurological and neuropsychological consequences of electrical and lightning shock: review and theories of causation. Neural Regen Res 12(5):677–686

Andrews CJ, Reisner A, Cooper MA (2017) Post electric or lightning injury syndrome: a proposal for a DSM formulation with implications for treatment. Neural Regen Res 12(9):1405–1412

Chudasama S, Goverman J, Donaldson J et al (2010) Does voltage predict return to work and neuropsychiatric sequelae following electrical burn injury? Ann Plastic Surg 64(5):552–555

Cooper MA (2001) Disability, not death, is the main issue. Natl Weather Dig 25:43

Cooper MA, Marshburn S (2005) Lightning strike and electric shock survivors, international. NeuroRehabilitation 20(1):43–47

Cooper MA, Marshburn S, Marshburn J (2001) Lightning Strike and Electric Shock, Survivors International. Natl Weather Dig 25(1, 2):48–50

Morse MS, Berg JS, ten Wolde RL (2004) Diffuse electrical injury: a study of 89 subjects reporting long-term symptomatology that is remote to the theoretical current pathway. IEEE Trans Biomed Eng 51(8):1449–1459

Pliskin N, Ammar A, Fink J et al (2006) Neuropsychological changes following electrical injury. J Intl Neuropsychological Soc 12(01):17–23

Pliskin NH, Meyer GJ, Dolske MC et al (1994) Neuropsychiatric aspects of electrical injury. A review of neuropsychological research. Ann NY Acad Sci 720:219–223

Primeau M (2005) Neurorehabilitation of behavioral disorders following lightning and electrical trauma. NeuroRehabilitation 20(1):25–33

Primeau M, Engelstatter GH, Bares KK (1995) Behavioral consequences of lightning and electrical injury. Semin Neurol 15(3):279–285

Reisner AD (2006) A case of lightning injury with delayed-onset psychiatric and cognitive symptoms. Brain Inj 20(10):1093–1097

Eye Injuries

Andrews CJ (1995a) Structural changes after lightning strike, with special emphasis on special sense orifices as portals of entry. Semin Neurol 15(3):296–303

Bullock JD (1978) The blindness of Saint Paul. Ophthalmology 85(10):1044–1053

Bullock JD (1994) Was Saint Paul struck blind and converted by lightning? Surv Ophthalmol 39(2):151–160

Campo RV, Lewis RS (1984) Lightning-induced macular hole. Am J Ophthalmol 97(6):792–794

Cazabon S, Dabbs TR (2000) Lightning-induced cataract. Eye 14(Pt 6):903–904

Dinakaran S, Desai SP, Elsom DM (1998) Telephone-mediated lightning injury causing cataract. Injury 29(8):645–646

Dinakaran S, Desai SP, Elsom DM (2001) Lightning-induced cataract. Eye 15(Pt 5):690

Dinakaran S, Desai S, Elsom D (2002) Ophthalmic manifestations of lightning lightning strikes [and author's reply]. Surv Ophthalmol 47(3):292–293

Espaillat A, Janigian R Jr, To K (1999) Cataracts, bilateral macular holes, and rhegmatogenous retinal detachment induced by lightning. Am J Ophthalmol 127(2):216–217

Grover S, Goodwin J (1995) Lightning and electrical injuries: neuro-ophthalmologic aspects. Semin Neurol 15(4):335–341

Gupta A, Kaliaperumal S, Sengupta S et al (2006) Bilateral cataract following lightning injury. Eur J Ophthalmol 16(4):624–626

Hunt L (2000) Ocular injuries induced by lightning. Insight 25(2):59–60

Kleiter I, Luerding R, Diendorfer G et al (2007b) A lightning strike to the head causing a visual cortex defect with simple and complex visual hallucinations. J Neurol Neurosurg Psychiatry 78(4):423–426

Moon SJ, Kim JE, Han DP (2005) Lightning-induced maculopathy. Retina 25(3):380–382

Norman ME, Younge BR (1999) Association of high-dose intravenous methylprednisolone with reversal of blindness from lightning in two patients. Ophthalmology 106(4):743–745

Norman ME, Albertson D, Younge BR (2001) Ophthalmic manifestations of lightning strike. Surv Ophthalmol 46(1):19–24

Rao K, Rao L, Kamath A et al (2009) Bilateral macular hole secondary to remote lightning strike. Indian J Ophthalmol 57(6):470–472

Rivas-Aguino PJ, Garcia RA, Arevalo JF (2006) Bilateral macular cyst after lightning visualized with optical coherence tomography. Clin Exp Ophthalmol 34(9):893–894

Sommer LK, Lund-Andersen H (2004) Skin burn, bilateral iridocyclitis and amnesia following a lightning injury. Acta Ophthalmol Scand 82(5):596–598

Yi C, Liang Y, Jiexiong O et al (2001) Lightning-induced cataract and neuroretinopathy. Retina 21(5):526–528

Ear Injuries

Andrews CJ (1995b) Structural changes after lightning strike, with special emphasis on special sense orifices as portals of entry. Semin Neurol 15(3):296–303

Bellucci RJ (1983) Traumatic injuries of the middle ear. Otolaryngol Clin North Am 16(3):633–650

Bergstrom L, Neblett LW, Sando I et al (1974) The lightning-damaged ear. Arch Otolaryngol 100(2):117–121

Cankaya H, Egeli E, Evliyaoglu Z (2002) Hearing loss caused by lightning strike - case report and review of the literature. J Otolaryngol 31(3):181–183

Gluncic I, Roje Z, Gluncic V et al (2001) Ear injuries caused by lightning: report of 18 cases. J Laryngol Otol 115(1):4–8

Gordon MA, Silverstein H, Willcox TO et al (1995) Lightning injury of the tympanic membrane. Am J Otol 16(3):373–376

Jones DT, Ogren FP, Roh LH et al (1991) Lightning and its effects on the auditory system. Laryngoscope 101(8):830–834

Kristensen S, Tveteras K (1985) Lightning-induced acoustic rupture of the tympanic membrane: (a report of two cases). J Laryngol Otol 99(7):711–713

Mora-Magana I, Collado-Corona MA, Toral-Martinon R et al (1996) Acoustic trauma caused by lightning. Int J Pediatr Otorhinolaryngol 35(1):59–69

Myung N-S, Lee I-W, Goh E-K et al (2012) Cochlear implantation for severe sensorineural hearing loss caused by lightning. Amer J Otolaryngol 33(6):767–769

Offiah C, Heran M, Graeb D (2007) Lightning strike: a rare cause of bilateral ossicular disruption. Amer J Neuroradiol 28(5):974–975

Ogren FP, Edmunds AL (1995) Neuro-otologic findings in the lightning-injured patient. Semin Neurol 15(3):256–262
Redleaf MI, McCabe BF (1993) Lightning injury of the tympanic membrane. Ann Otol Rhinol Laryngol 102(11):867–869
Richmond DR, Yelverton JT, Fletcher ER et al (1989) Physical correlates of eardrum rupture. Ann Otol Rhinol Laryngol 140:35–41
Wright JW Jr, Silk KL (1974) Acoustic and vestibular defects in lightning survivors. Laryngoscope 84(8):1378–1387

Injuries to the Skin

Aslar AK, Soran A, Yildiz Y et al (2001) Epidemiology, morbidity, mortality and treatment of lightning injuries in a Turkish burns units. Int J Clin Pract 55(8):502–504
Cherington M, Olson S, Yarnell PR (2003) Lightning and Lichtenberg figures. Injury 34(5):367–371
Cohen MA (2001) Clinical pearls: struck by lightning. Cutaneous manifestation of lightning strike (splash). Acad Emerg Med 8(9):893, 929–893, 931
Domart Y, Garet E (2000) Images in clinical medicine. Lichtenberg figures due to a lightning strike. New England J Med 343(21):1536
Holstege CP (2005) Images in emergency medicine. Lightning strike. Ann Emerg Med 45(4):354–362
Ikpeme IA, Udosen AM, Asuquo ME et al (2007) Lightning burns and traditional medical treatment: a case report. West Afr J Med 26(1):53–54
Jonas L, Fulda G, Nizze H et al (2002) Detection of gold particles in the neck skin after lightning stroke with evaporation of an ornamental chain. Ultrastruct Pathol 26(3):153–159
Kumar V (2007) Filigree burn of lightning: two case reports. Med Sci Law 47(2):171–173
Mahajan AL, Rajan R, Regan PJ (2008) Lichtenberg figures: cutaneous manifestation of phone electrocution from lightning. J Plast Reconstr Aesthet Surg 61(1):111–113
Matthews MS, Fahey AL (1997) Plastic surgical considerations in lightning injuries. Ann Plast Surg 39(6):561–565
McKrady-Kahn VL, Kahn A (1981) Lightning burns. West J Med 134(v3):215–219
Muehlberger T, Vogt PM, Munster AM (2001) The long-term consequences of lightning injuries. Burns 27(8):829–833
Murty OP (2006) Dramatic lightning injury with exit wound. J Forensic Legal Med 14(4):225–227
Murty OP (2009) Lightning fatality with blast, flame, heat and current effects: a macroscopic and microscopic view. J Forensic Legal Med 16(3):162–167
Resnik BI, Wetli CV (1996) Lichtenberg figures. Am J Forensic Med Pathol 17(2):99–102
Ritenour AE, Morton MJ, McManus JG et al (2008) Lightning injury: a review. Burns 34(5): 585–594
Selvaggi G, Monstrey S, Van Landuyt K et al (2005) Rehabilitation of burn injured patients following lightning and electrical trauma. NeuroRehabilitation 20(1):35–42

Reproductive System: Pregnancy

Garcia Gutierrez JJ, Melendez J, Torrero JV et al (2005) Lightning injuries in a pregnant woman: a case report and review of the literature. Burns 31(8):1045–1049
Rees WD (1965) Pregnant woman struck by lightning. Br Med J 1(5427):103–104

Chapter 4
Research on Pathophysiology of Medical Effects by Lightning

Abstract Explanations of the pathophysiology of lightning injuries are largely speculative, not based on research. Worse, sometimes they are extrapolated from research on electrical injuries, another injury entirely, particularly in regard to tissue injury. Research is difficult due to the difficulty of producing inexpensive, replicable, and valid "lightning," the difficulties of developing an animal model, the infrequency of victims making case series studies difficult, and the lack of funding. Since the pathophysiology is largely unknown and animal models are incomplete, little prospective controlled research, as is done on most other mechanisms of trauma and disease, has been done. Therapy is largely empirical – based on experience, best guesswork, and application of studies for similar types of trauma, not derived from research or evidence-based care of lightning strike survivors. This chapter will discuss foundational animal research, list research questions that remain unanswered, and discuss some of the difficulties and decisions that must be made for any laboratory or medical research project on lightning injuries.

4.1 Introduction

Ideally, pathophysiology guides therapy. For the lightning injuries that we cannot prevent, physicians should know how to treat the survivor based on known and certain pathophysiology using research and evidence-based therapies. To be fair, we should say that the pathophysiology of other traumas is not always well known either.

There is no lack of lightning pathophysiology questions to answer (see sidebars and lists throughout the chapter), but few of them have been investigated. Papers on lightning injury pathophysiology have almost always been more educated speculation than fact supported by research. There are more questions (Table 4.1) than answers in this chapter and in lightning injury – but that offers opportunities for the new researcher!

Lightning injuries can be described and even quantified to some extent, but the treatment is still largely empirical, not based on research or evidence-based medicine. To be fair, most medical care, with the exception of immunizations, has been

M. A. Cooper, R. L. Holle, *Reducing Lightning Injuries Worldwide*,
Springer Natural Hazards, https://doi.org/10.1007/978-3-319-77563-0_4

Table 4.1 Areas to explore in lightning injury

Blunt trauma – Explosive injury (Blumenthal 2016)
Macroscopic/structural changes – Direct damage
Contusion
Cellular damage
Pathway of the injury
Orifice entry
Flashover – How much goes through versus around
Neurochemical changes
ANS effects
Electrical effects
Electroporation
Cellular-level mechanical effects
Cellular-level enzymatic effects
Subcellular organelle damage

Empiric therapy is therapy based on experience and, more specifically, *therapy* begun on the basis of a clinical educated guess in the absence of complete or perfect information (Wikipedia).

largely empiric and has only been based on known pathophysiology in the last century or so.

Pathophysiology is more likely to be known for illnesses caused by bacteria or viruses than for trauma. What actually happens to the neurons, synapses, chemical end plate receptors, and other microscopic and biochemical components of brain function in brain injury, whether caused by lightning, automobile accidents, sports concussion, or an improvised explosive device? What is responsible for the loss of executive function, the personality changes, and the difficulty with processing all parts of memory (Andrews 2017)? Neuropsychological testing can help to delineate cognitive and functional injury and tailor therapy, but it is not sensitive nor reliable enough for many observers, including the courts, when there is a question of whether the injury occurred, as claimed, or what the extent of the injuries and prognosis will be. Legally, having a "picture" or research literature-based description of the injury is often necessary to serve as diagnostic "proof" in court. Fortunately, fMRI, PET scans, and other imaging techniques show some promise in helping to define the lesions, lead to more focused care, and guide research into the pathophysiology of the injuries. What is still necessary for these tests to be able to explain or guide research into the pathophysiology? Will knowing the pathophysiology help guide therapy and treatment?

4.2 Pathophysiology of an Injury

4.2.1 Injury Cascade

We are all familiar with the body's response to common insults. Most of us could describe the injury cascade, the sequence of development of signs and symptoms, that occurs with sunburn, poison ivy, a sprained ankle, or a broken bone that all have relatively predictable changes over minutes to hours to weeks to sometimes months. A cascade of effects is probably started by lightning injury, some of which we can describe or theorize about from clinical observation (Table 4.2). However, it has not been defined nor even demonstrated in either human subjects or animals. More importantly, it is unknown if anything can be done to interrupt the cascade of effects set in motion by the injury.

We know that exposure to lightning is nearly instantaneous, too short to cause significant burns in most people. We know clinically that the nervous system is the most vulnerable but cannot tell how nerve cells and synaptic connections are injured or what determines whether they will recover or die immediately, in a few hours or

Table 4.2 Injury cascade: hypothesized injury timeline

The electrical insult is different than high- or low-voltage electricity
Very short, very fast
Flashover
Exposure often too short to cause burns
What are the electrical field effects to cells?
"Charging of cells"
Electroporation
Subcellular damage
Acute injury (Chap. 3)
Cell death – immediate or delayed?
What is the process of cell repair/recovery, and how does it differ with tissue type?
Late onset of some symptoms
Sleep disturbance
Pain
Seizures
How much can technology help to define injury?
CT, MRI, EEG, EMG, fMRI, PET
Is there increased predisposition to other illnesses?

days or after prolonged efforts to heal themselves. We do not know if cellular changes are responsible for the brain injury that we observe. We do not know if "scar" tissue from the dead cells develops nor how long it takes to form. We do not know if that is what is responsible for the delayed seizures and other symptoms that sometimes occur after weeks to months. We also do not know if lightning injury predisposes the survivor to premature arthritis, cancers, diabetes, and other endocrine, rheumatologic, or immune diseases. To do this, a sufficiently large number of survivors would need to be monitored for years to see if they develop illnesses or other complications statistically out of proportion to what would be expected compared to an uninjured control population matched by age, sex, and other risk factors.

There is often a well-known and predictable history of what happens to tissue after an injury. In high-velocity gunshot wounds, there is often cavitation and significant tissue injury surrounding the pathway of the bullet as well as fragmentation of the missile; in myocardial infarction (heart attack), there will be deterioration of the cells injured often causing arrhythmias and changes to the pumping efficiency.

Lightning injury is no different. When the injury is triggered, a cascade of effects occurs over hours and days after the injury. Some have been described, and others theorized, but all need more definition and quantification (Table 4.2).

4.2.2 What Affects the Acute Lightning Injury?

Although the mechanisms of injury have been studied and well characterized, it is unknown what other factors may determine the extent of an injury. Some potential factors are listed in Table 4.3.

It is reasonable to expect that lightning injuries might be different depending on different characteristics of the lightning insult such as positive versus negative lightning, the victim's proximity to the strike and mechanism of injury, number of return strokes, as well as other conditions surrounding the incident such as temperature, wind, and exposure of the victim. Knowing these characteristics may help with the epidemiology or demographics of lightning but are of little or no help in preventing or modifying an injury.

It has been theorized that the likelihood of a lightning injury to be fatal may be related to the timing of the lightning strike, particularly if it occurs during a more vulnerable portion of the cardiac cycle. Cooper's animal work set out to investigate this possibility but, due to insufficient funding, was unable to gather more than preliminary conclusions about the effect on the autonomic nervous system (Kotsos et al. 1998).

Common imaging techniques, computerized tomography (CT) and magnetic resonance imaging (MRI), are "anatomic" tests that show only a picture, not a functional evaluation of how the brain, heart, nerves, or other structures work normally, much less when injured. Although Lee has corroborated electroporation of muscle cell walls with high-voltage electrical injury, the physics of high-voltage electrical

Table 4.3 What other factors may affect lightning injury?

Lightning characteristics
"Type" of lightning
Positive versus negative versus long continuous current (LCC) lightning
Mechanism of injury
"Dosage" of lightning delivered
Timing of hit compared to the cardiac cycle
Other meteorological conditions – Wind, cold, and other factors?
Where people are/what they were doing?
Blunt trauma – concussive injury
Structural changes – direct damage
Pathway of the injury
Orifice entry (Andrews 1995)
Flashover – how much goes through versus around
Neurochemical changes
Autonomic nervous system effects
Electrical effects
Electroporation
Cellular-level mechanical effects
Cellular-level enzymatic effects
Subcellular organelle damage

injury is quite different than lightning (Lee et al. 1995, 1997). Electroporation, the formation of pores in the cell walls due to electric field effects, has never been investigated in the lightning model nor investigated in nerve or brain cells that are more affected clinically by lightning than muscle tissue.

4.2.3 Can the Injury Cascade Be Interrupted?

As a rule, nervous system injury, once it has occurred, is largely irrecoverable. Most therapies are based on rebuilding skills, movement, and brain function by retraining other tissues and nerve pathways, using braces, medications to mitigate seizures, pain, and other sequelae, cognitive therapy, and other modalities.

Even if the pathophysiology becomes known, it may not be possible to design a treatment to prevent or mitigate CNS and other system injury from lightning. If there were an effective CNS-protective agent, there is no likelihood that a person could take the drug prior to the injury. On the other hand, knowing the pathophysiology may help us to discover better ways of treating the injury once it has occurred.

If it were possible to collect a sufficiently large study group, utilizing the therapies that have been found to be useful for post-concussive injury, for instance, might be tried for lightning CNS injury as well. Lacking a study group, therapy for an individual becomes empiric and hopeful. It may work, or it may not.

Research and Therapy Questions

How much will knowing the pathophysiology help us in caring for patients? Will it improve the care that we can give and improve recovery? How soon after the lightning strike does the cascade of central nervous system (CNS) injury that leads to cognitive disability and the other sequelae of lightning injury start? How likely are interventions that interrupt this cascade at various points to improve recovery? How similar is the CNS injury from lightning to CNS injury or postischemic injury from other injuries such as concussion or stroke/cerebrovascular accident?

4.3 Research Funding for Medical Studies

While there have been attempts at modeling lightning injury and replicating what happens to humans using animals (Kitagawa et al. 1986, 1990, 2001; Andrews et al. 1989, Cooper and Kotsos 1997), to date, these remain isolated and mostly unfunded. In the United States, research funding generally goes to the most pressing issues affecting the population such as cancer, heart disease, bioterrorism, and HIV. This is in large part due to the fact that these have the largest affected populations and, consequently, the most extensive lobbying efforts by support groups and politicians. "Orphan" diseases and injuries have almost no funding except through private foundations concerned with the particular problem.

While many Americans might expect that the National Weather Service in NOAA would be an ideal place to find funding for the medical effects of lightning, these agencies are often among those with the least government funding for the extensive forecasting and warning work they are tasked to do, much less to fund medical research. The result is that the pathophysiology of lightning injury is largely unstudied because it is unfunded, resulting in treatment and rehabilitation that is not based on organized, controlled, prospective studies.

Unfortunately, because lightning is so expensive and difficult to replicate in the laboratory and because so few people are injured or killed in most countries, it is unlikely that research funding to determine the pathophysiology will be a priority for most governments or research institutions.

4.4 Laboratory and Clinical Studies

4.4.1 *Early Studies*

A tragic incident occurred in 1967 where 11 of 46 high school students hiking through the Japanese Alps were killed and 14 seriously injured by lightning. Following this event, Nobu Kitagawa, Ph.D., a physicist specializing in lightning phenomenology in Japan, partnered with high-voltage engineers and physicians, some of whom had given medical care to the students, to investigate the mechanisms of injury. Over the next 35 years, he and his associates answered many fundamental questions about lightning injury, designing experiments using dummies, mice, rats, and rabbits in high-voltage laboratories where lightning-like pulses could be generated. They also investigated numerous lightning incidents in the field for a nearly 40-year period. Based on their findings, they wrote some of the first lightning safety guidelines (Andrews et al. 1996).

A selection of the questions answered in the studies by Kitagawa et al. include the following results:

1. The median value of fatal threshold energy of lightning was weight dependent and was 62.58 +/− 11.93 J/kg for rats, mice, and rabbits (Ohashi et al. 1986, 2001).
2. Artificial respiratory support given to shocked rabbits increased survival by over 25% (Ishikawa et al. 1981).
3. In an attempt to investigate whether return strokes had a summative or isolated effect, they designed an apparatus to give three strokes, each with a different energy level. They found that the lethal threshold of the energy in the rabbit depended on the maximum value of the energy of an individual dose, not on the sum of the energy of the three successive voltage impulses (Nagai et al. 1982).
4. The faster that flashover occurred, the more likely the person or animal was to survive, concluded from the study of 140 people injured over 17 years as well as laboratory work on mice and rats (Ohashi et al. 1986).
5. Using two dummies, one "control" and one with interventions such as a raincoat, boots, or metal on their head, with the HV lightning electrode placed equidistantly between their heads. The dummies had coatings of paint that replicated the 300–500 ohm resistance of humans. They found that:

    (a) The presence of either vinyl raincoats or rubber boots on the dummies or of metal around the head had no effect on the lightning distribution between the dummies. They offered no protection.
    (b) Strikes were directly related to the height of an object such as a golf club, umbrella, or wooden pole held above the head, not to the object's material. It always hit the highest point of the object. In other words, holding something over your head will increase the chance of being hit, regardless of whether the object is made of metal, plastic, or wood.

(c) The skin exerts no insulating effect, and the human body responds as a conductive body of 300–500 ohms from head to both feet.
(d) When the amount of conduction current per body weight going through the body exceeds a certain level, it causes respiratory and cardiac arrest, resulting in death.
(e) Metal pieces on the body trigger and enhance surface discharges and surface flashovers and tend to reduce the fatal (inner) conduction current.
(f) The surface discharge on the body occurs at the voltage gradient, which is about one-half of the spark voltage gradient of the air. A stroke on the body involves two or three of the following stages:

    (i) When the lightning current is very low, the whole current flows through the body as conduction current.
    (ii) When the current increases to a certain level, surface discharges develop on the body.
    (iii) When the current increases still higher, the surface flashover bridges the strike point and the ground. In this stage, most of the lightning current flows as an arc current through the air and, only a very low current fraction flows through the body (flashover effect).

(g) No metal pieces on the body, but the human body itself protruding from the ground, are responsible for a lightning flash on the body. In case of an additional object, such as an umbrella, a golf club, or a fishing rod protruding higher than the body, the inducing effect is amplified, the effect being dependent not on the conductivity but its position to the body and how the height is increased.
(h) The vicinity of tall objects such as trees, masts, and chimneys, which are not equipped with lightning conductors, is very dangerous for two reasons: first, they are superior targets for lightning; second, it is highly probable that the main lightning current will flow onto a nearby human body as a sideflash when the tall objects are struck by lightning (Kitagawa et al. 1986).

6. From the clinical study of lightning victims, it was concluded that step voltage could be divided into two mechanisms: ground current through the surface of the earth caused little damage, whereas ground arcing caused much more serious injury (Kitagawa et al. 2001).
7. From studying the records of 256 victims between 1965 and 1999, it was suspected that skull fracture, intracerebral hemorrhage, solid organ rupture, and pulmonary contusion, though uncommon in lightning injury, were the results of concussive injury from vapor explosion of water on body surface during flashover and corroborated this in the laboratory by producing these injuries in small animals in frequent collaboration with Ishikawa and Kitagawa (Ohashi et al. 2001).

4.4.2 Other Laboratory Studies

As part of his dissertation, Christopher Andrews of Australia studied multiple aspects of lightning injury, replicating Cooper (1980), performed the first organized study of the long-term effects of lightning using survivors of telephone-mediated lightning injury, and studied lightning injury in the laboratory using sheep (Andrews and Darveniza 1989a, b; Andrews et al. 1989: Andrews 1995). Later, Andrews also noticed that lightning caused an increased QT interval on human ECG's post-strike and posited this as a causative factor in cardiac death from lightning from Torsades de Pointes (Andrews and Colquhoun 1993).

In the mid-1990s, with collaboration from Kitagawa and Andrews, Cooper set up an animal model of lightning injury utilizing variant hairless rats and a tabletop lightning generator. Cooper was able to replicate all of the human clinical findings of lightning injury but had to stop work due to lack of research funding before a standardized lightning "dose" was identified that could be utilized in controlled studies (Cooper and Kotsos 1997; Cooper et al. 2001a, b).

4.4.3 Other Clinical Studies

Another major researcher in lightning studies is Michael Cherington, M.D., of Colorado in the United States. He published multiple papers on the classification and rehabilitation of lightning injuries, along with multiple case studies (Cherington et al. 1992, 1993; Cherington 1995, 2003, 2005a, b).

4.5 Difficulties with Lightning Research

There are many levels at which research into the pathophysiology of lightning could be investigated. Let us systematically look at the pros and cons of each in Table 4.4.

4.5.1 Research Using Humans

In most countries, there are no easy ways to find and recruit people who have been injured or killed by lightning to make up a study. There are many reasons for this situation:

1. While death certificates may code for lightning injury in some countries, few, if any, countries require reporting of lightning injury for survivors, so there are no easily accessible collections. Many investigators have used newspaper reports for gathering statistics, but recruiting study participants this way is likely to be

Table 4.4 Challenges with lightning research

Human research
Recruitment of cases
Study biases
Dispersion of subjects
Cases ideally should be free from:
Diabetes and other neuropathic illnesses
Drug abuse history
Psychiatric history
Blunt head trauma
Animal research
Expensive – Both animals and equipment
Large number of animals required for some studies
Difficult signal processing problems
Monitoring equipment design
Shock timing control considerations
Definition of "dose"
Standardization of "dose"
Flashover effect
Molecular biology
Cell culture, blood levels, etc.
Requires specialized techniques
Bioengineering, collaboration
May be expensive

difficult as the contact information may never have been gathered or may be unavailable due to legal or privacy issues.
2. Many people who are injured by lightning do not go to a hospital initially, so they will not be entered into any medical or hospital databases.
3. People in rural areas or developing countries may not seek medical care due to distance or expense.
4. The only known collection of survivors is a support group in the United States (Cooper et al. 2001a, b). A few studies have been done using this population but are limited in several ways, including the bias associated with people who seek to join a support group compared to those who do not.

Prospective studies are naturally impossible as no researcher would put volunteers out in a thunderstorm to be struck and then studied for the results. Most of the medical literature on lightning injuries is made up of case reports, the most limited form of "research reporting." Nevertheless, retrospective review of cases has led to some useful information (Cooper 1980; Andrews and Darveniza 1989). Reviews done by the Kitagawa group produced several questions that the group took to the laboratory to investigate (Sect. 4.4.1).

If a researcher could collect a population, studying acute injury would require relatively recently injured patients (no more than 6–12 months), optimally with a "clean" history. That is, ideally, they should be free of illnesses that frequently affect the nervous system such as diabetes, hypertension, depression, neuropathic pain, alcohol or drug abuse, or prior brain injury. If only the electrical nature were to be studied, it would be desirable to assure that the subjects were free of concussive injury which sometimes accompanies lightning injury (see Mechanisms of Injury, Chap. 3).

Most studies of the long-term consequence of lightning strike have been descriptive, such as those associated with the corded telephone (Andrews and Darveniza 1989b; Cooper 2001). For more sophisticated questions, such as whether early dementia, cancers or cardiovascular disease are more likely after lightning injury, as some survivors claim, a large cohort of survivors would need to be studied over an extended time period, as well as involve control groups of matched non-lightning injured individuals, a very expensive type of study. Short of this effort, a survey of survivors after many years could be done with comparison to known statistics on these types of illnesses. Even these surveys would be hampered by national dispersion of survivors, inability to objectively validate lightning strike or mechanism in some cases, response rate, and other methodological issues.

While all of these caveats could discourage research, if the researcher is motivated, many of these can be overcome or taken into account. However, nearly any data honestly obtained are better than no data.

4.5.2 *Animal Research*

Animal work is a reasonable alternative. Kitagawa, Ishikawa, Ohashi, Andrews, and Cooper have all done very credible research using animals (Ohashi et al. 1986; Andrews 1995). Cooper has built a reasonable rat model, inducing most of the clinical signs that are seen in humans (Cooper 2002). Unfortunately, such research is expensive, time-consuming, and sometimes requires large numbers of animals to show significant differences.

Many other hurdles must be overcome in research equipment design, anesthesia choice, and animal care. Mechanical factors, such as the size or type of switches to avert dielectric breakdown across parts of the equipment, can be a problem. For instance, any type of monitoring or connection to sensitive equipment must be discontinued during a shock, or the energy will travel preferentially along the monitor wires and blow out the equipment instead of shocking the animal. Measurement of specific parameters may require sophisticated software and expensive equipment. If the portion of the cardiac cycle that is shocked is the study question, it is obviously necessary to shock at the same portion of the cardiac cycle on each animal, but, because the animal will be unhooked from the computer and monitoring equipment just prior to and during the shock, the precise strike time would need to be calculated to predict the appropriate time to shock the animal. With a heart rate of 200–300 or more beats/minute for small animals, this can be difficult and requires special computer hardware and software programming.

Some anesthetics are neuroprotective, while others affect cardiac function. Temperature control in anesthetized rats is a significant issue. Animals that are being studied for neurocognitive injury need a standardized, quiet environment, which is not always possible in the cramped quarters of most animal facilities.

While both Kitagawa's group, Cooper, and Andrews were all able to replicate certain aspects of lightning injury in animals, doing controlled studies requires a standard "dose" of lightning to be delivered to each animal. This is difficult, given flashover. In the Kitagawa group's early experiments, they implanted the delivery electrode in the animal's scalp, intubated them, and immersed them in oil to prevent flashover. However, the researchers were never able to standardize this procedure (personal communication). Kitagawa's group found that the median value of fatal threshold energy for lightning was weight dependent and was 62.58+/− 11.93 J/kg for rats, mice, and rabbits (Ohashi et al. 1986, 2001). Their experiments were conducted in their high-voltage laboratory, but few of these are available around the world, and the setup for individual experiments can be quite expensive. Given this difficulty, is a high-voltage (expensive) laboratory necessary? Can a less expensive per-shock tabletop apparatus, similar to the one Cooper developed, deliver adequate energy, properly wave shaped and compatible with other lightning parameters (Fig. 4.1)? Further, is the dose for small animals applicable (scalable) to humans?

Almost all lightning studies reported in this chapter were done in the dark so that photography could be used to document pathway and other parameters. With the

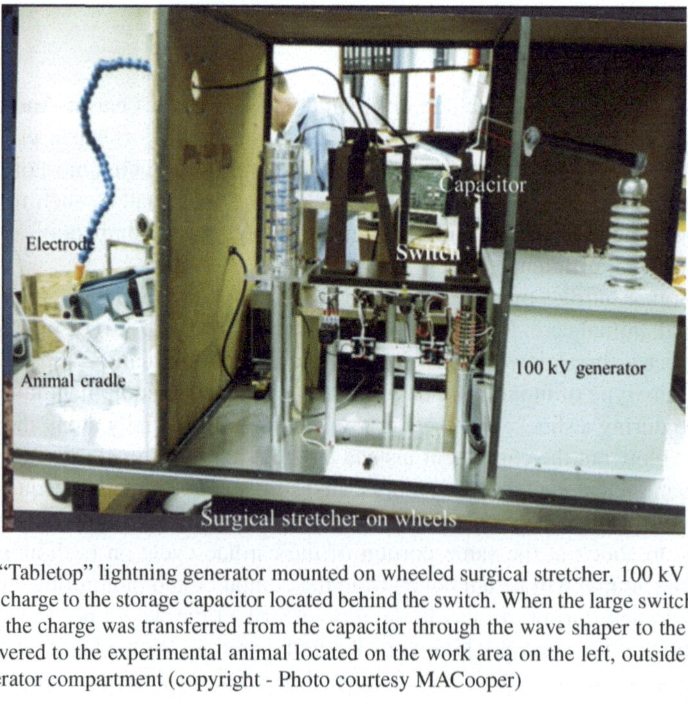

Fig. 4.1 "Tabletop" lightning generator mounted on wheeled surgical stretcher. 100 kV generator delivered charge to the storage capacitor located behind the switch. When the large switch arm was activated, the charge was transferred from the capacitor through the wave shaper to the electrode to be delivered to the experimental animal located on the work area on the left, outside the lightning generator compartment (copyright - Photo courtesy MACooper)

advent of very fast frame speeds (millions/second), this would be an exciting, but expensive, tool to employ.

4.5.3 *Molecular Biological Research*

While it is possible that tissue culture or biological experiments might be appropriate rather than using a "whole-animal preparation," to date, no research designs for these techniques have been developed, and most researchers currently involved with lightning have neither the expertise nor funding to recruit collaborators to begin this line of inquiry. The Kitagawa group did not study cellular damage in the animals they used. While Andrews looked at the brain tissue, both he and Cooper found it difficult to find interested animal pathologists to assist in their research by looking at the microscopic tissue damage level, much less to study how cellular function or cerebral connections might be affected. The sophisticated and currently available clinical and research imaging used for humans was not available for small animal work at the time of their studies.

4.5.4 *Other Difficulties with Research Design*

The pathophysiology of lightning injury may never be known because of the difficulty in doing research with lightning on living tissue or even on nonliving tissue, for the following reasons:

1. To do research on lightning, the mechanism of injury must replicate lightning as closely as possible. It is not acceptable to use generated electricity in ultrashort bursts and call it "lightning." Proper knowledge of lightning-appropriate wave shapes and other physical properties must be applied.
2. Lightning, with its true parameters, is expensive to simulate/replicate in the laboratory. The high-voltage laboratory in which it occurs is not likely to be located near animal facilities where in vivo experiments can be conducted.
3. Ideally, a multidisciplinary team of lightning engineers, veterinary pathologists, physiologists, and others would be needed to plan and investigate many of the questions we have posed in this chapter. Gathering a team with the proper background, curiosity, time, and funding is a challenge.

4.6 Selected Research Questions

Table 4.5 lists a number of specific questions that could be pursued in medical research and perspective on their related phenomena.

Table 4.5 Selected research questions on lightning effects on the human body

Cardiac injury
Is cardiac arrest from lightning caused by:
Direct injury to the heart?
Thrombosis of vessels
Vascular spasm
Direct tissue injury
Injury to control mechanisms?
Damage to cardiac and respiratory centers in the brain.
Damage to pacemakers (carotid, AV node, others) – Some of these are quite close to the body surface. It is unknown if the intense, nearly instantaneous flow of lightning current flashover could induce electrical effects in the pacemakers to cause arrhythmias or arrest.
Damage to the electrical signal conducting system.
Damage to autonomic nervous system controls.
Timing of the strike during the "vulnerable period" of the cardiac cycle?
Other causes?
Nervous system injury
1. Is there a simple and inexpensive way to prevent keraunoparalysis so that those sleeping in thatched buildings could escape from the flames in developing countries, and how would you test this?
2. Are the nervous system and subsequent neuropsychological/behavioral changes due to:
Chemical/cellular brain changes - how, where? Is there a way to reverse them or to stop the progression of the injury pathophysiology?
Synaptic disruption, damage to pathways through the corpus callosum, or other structures that normally facilitate linking of brain messages to accomplish a task?
Blunt injury from the head being jerked around from induced muscle spasm or concussive injury of the lightning blast wave? Does it differ from other post-concussive syndromes?
Damage to the hippocampus/hypothalamus or other structures to account for the memory processing difficulties?
Damage to the frontal lobes to explain emotional control deficits?
Skin injury
Are these primary burns from beta particle injury as hypothesized by Cooray et al. (2015)?
Are they secondary injury as flashover energy turns rain or sweat on the skin into steam that burns?
What is the mechanism for burns from metal close to the body? Heat? Secondary electrical discharge from the charged metal? How would you prove this?
Why don't more victims have skin marks/burns?
What is the mechanism for Lichtenberg figures? Do they follow blood vessel or nerve distributions? Why do they resemble fractals? One theory is that lightning energy coursing over the body causes superficial capillaries to constrict, forcing red cells through the vessel walls that resembles a bruise type of injury. How would you prove/disprove this? What other hypotheses could you propose?
There is good anecdotal evidence from several reports that burns can be much more severe in developing countries, not because of the lightning but as a result of secondary injury from burning thatch falling on victims while they have acute keraunoparalysis. Are there other injuries that are different in developing countries vs developed countries? Which ones and why? Reference "tiptoe sign" in South Africa.

(continued)

Table 4.5 (continued)

Mechanisms of injury
1. Are there other mechanisms of injury yet to be defined?
2. Is there medical evidence, beyond mathematical and engineering conjecture, that direct strikes are more often fatal than other mechanisms of injury or that the degree of injury is worse? How would you collect this data? Is it possible to design a research project that would answer this? Would this be in the lab or a clinical study?
3. Is the distribution of lightning injuries different in developing countries than developed countries? If yes, how and why?
4. Is the distribution of injury mechanisms (Chap. 2) different in developing countries?

References

Andrews CJ (1995) Structural changes after lightning strike, with special emphasis on special sense orifices as portals of entry. Semin Neurol 15(3):296–303

Andrews CJ, Colquhoun D (1993) The QT interval in lightning injury with implications for the 'cessation of metabolism' hypothesis. J Wildernz Med 4:155–166

Andrews CJ, Darveniza M (1989a) Effects of lightning on mammalian tissue. In: Preprints of the international conference on lightning and static electricity, Bath, United Kingdom, 26–28 Sept 1989

Andrews CJ, Darveniza M (1989b) Telephone mediated lightning injury – an Australian survey. J Trauma 29(5):665–671

Andrews CJ, Berger G, Floret N et al (1996) International safety and rescue guide against lightning hazards. In: Proceedings of the 23rd international conference on lightning protection, Florence, Italy, 23–27 Sept 1996

Andrews CJ, Darveniza M, Mackerras D (1989) Lightning injury: a review of clinical aspects, pathophysiology and treatment. Adv Trauma 4:241–287

Andrews CJ, Reisner A, Cooper MA (2017) Post electric or lightning injury syndrome: a proposal for a DSM formulation with implications for treatment. Neural Regen Res 12(9):1405–1412

Blumenthal R (2016) The explosive effects of lightning: what are the risks? Acad Forensic Pathol 6:89–95

Cherington M (1995) Central nervous system complications of lightning and electrical injuries. Semin Neurol 15(3):233–240

Cherington M (2003) Neurologic manifestations of lightning strikes. Neurology 60(2):182–185

Cherington M (2005a) Neurorehabilitation of the multifaceted and complicated neurologic problems associated with lightning and electrical injuries. NeuroRehabilitation 20(1):1–2

Cherington M (2005b) Spectrum of neurologic complications of lightning injuries. NeuroRehabilitation 20(1):3–8

Cherington M, Yarnell P, Hallmark D (1993) MRI in lightning encephalopathy. Neurology 43(7):1437–1438

Cooper MA, Marshburn S, Marshburn J (2001b) Lightning strike and electric shock, survivors international. Natl Weather Dig 25(1, 2):48–50

Cherington M, Yarnell P, Lammereste D (1992) Lightning strikes: nature of neurological damage in patients evaluated in hospital emergency departments. Ann Emerg Med 21(5):575–578

Cooper MA (1980) Lightning injuries: prognostic signs for death. Ann Emerg Med 9(3):134–138

Cooper MA (2001) Disability, not death, is the main issue. Natl Weather Dig 25:43

Cooper MA (2002) A fifth mechanism of lightning injury. Acad Emerg Med 9(2):172–174

Cooper MA, Kotsos TP (1997) Development of an animal model of lightning injury with flashover utilizing a table-top lightning generator. In: Proceedings of lightning and mountains'97, Chamonix Mont-Blanc, France, 1–5 June 1997

Cooper MA, Kotsos T, Gahndi MV et al (2001a) Acute autonomic and cardiac effects of simulated lightning strike in rodents. Soc Acad Emerg Med (Atlanta, Georgia)

Cooray V, Cooray GK, Cooray C (2015) On the possible mechanism of keraunographic markings on lightning victims. J Atmos Sol Terr Phys 136:119–123

Ishikawa T, Miyazawa T, Ohashi M et al (1981) Experimental studies on the effect of artificial respiration after lightning accidents. Res Exp Med 179:59–68

Kitagawa N, Ohashi M, Ishikawa T (1990) Safety guide against lightning hazards. Res Lett Atmos Electricity 10:37–44

Kitagawa N, Ohashi M, Ishikawa T (2001) The substantial mechanisms of step voltage effects. Atmos Electricity 21(2):87–94

Kitagawa N, Tsurumi S, Ishikawa T et al (1986) The nature of lightning discharges on human bodies and the basis for safety and protection. In: Preprints of the 18th international conference on lightning protection, Munich, Germany, 16–20 Sept 1986

Kotsos TP, Zimmermann S, Cooper M (1998) Acute cardio-respiratory effects of simulated lightning pulse applied across hairless rats. In: Preprints of the 15th international lightning detection conference, Tucson, Arizona, 16–18 Nov 1998

Lee RC (1995) Biophysical mechanisms of cell membrane damage in electrical shock. Semin Neurol 15:367

Lee RC (1997) Injury by electrical forces: pathophysiology, manifestations, and therapy. Curr Probl Surg 34(9):677–764

Nagai Y, Ishikawa T, Ohashi M et al (1982) Study of lethal effects of multiple stroke flash. Lethal effects on rabbits subjected to 3-successive voltage impulses simulating multiple stroke flash. Res Lett Atmos Electricity 2:87–90

Ohashi M, Kitagawa N, Ishikawa T (1986) Lightning injury caused by discharges accompanying flashovers – a clinical and experimental study of death and survival. Burns Incl Therm Inj 12(7):496–501

Ohashi M, Hosoda Y, Fujishiro Y et al (2001) Lightning injury as a blast injury of skull, brain, and visceral lesions: clinical and experimental evidences. Keio J Med 50:257–262

Chapter 5
Economic Damages of Lightning

Abstract The economic effects of lightning damage to property are large, varied, and widely spread across society. In addition to loss of life and health (Part II), there are direct and indirect costs from lightning which affect a wide spectrum from individual homeowners and small businesses to large companies such as oil refineries as well as national heritage sites. Unfortunately, even in developed countries, little systematic research has been done on these areas, and research that has been completed has often been considered proprietary by industry or insurance companies. There are few models to predict losses and no routine monitors to measure them. As a result, figures for economic damages may be fraught with reporting errors.

5.1 Personal and Small Business Property Losses: The United States

Personal and small business properties suffer lightning damage both from direct damage and indirectly when current is transmitted into the structures through plumbing or wiring that is hit by lightning some distance away. In one of the first studies of lightning property damage claims in the United States (Holle et al. 1996), reports from contributing insurance companies covering three western states for a period of 5 years from 1987 to 1991 showed homeowner and business claims totaling 5755 cases annually with an average of $916 per claim. At that time, homeowner claims were reported 11 times more frequently than those of small businesses. When the authors extrapolated their study to include all insured homes and businesses across the United States, they calculated 307,000 claims totaling $332,000,000 (Holle et al. 1996).

Other reports have been sparse until recently. Archived reports from the Insurance Information Institute and State Farm® go back to 2005 (http://www.iii.org/table-archive/20504). Table 5.1 shows their reported data on the number of homeowner claims from 2011 through 2015. In 2015, the number of homeowner insurance claims from lightning strikes in the United States, just under 100,000, totaled $790

M. A. Cooper, R. L. Holle, *Reducing Lightning Injuries Worldwide*, Springer Natural Hazards, https://doi.org/10.1007/978-3-319-77563-0_5

Table 5.1 Homeowners' insurance claims and payout for lightning losses, 2011–2015

						Percent change	
	2011	2012	2013	2014	2015	2014–2015	2011–2015
Number of paid claims	186,307	151,000	114,740	99,871	99,423	**−0.4%**	**−46.6%**
Insured losses ($ millions)	$952.5	$969.0	$673.5	739.0	739.0	**+6.9%**	**−17.0%**
Ave. cost per claim	$5112	$6400	$5869	$7400	$7947	**+7.4%**	**+55.5%**

Source: Insurance Information Institute, State Farm® website http://www.iii.org/fact-statistic/lightning

Table 5.2 Top ten states in the United States for homeowners insurance lightning losses by number of claims in 2015

Rank	State	Number of paid claims	Insured losses ($millions)	Average cost/claim
1	Florida	11,898	$156.2	$13,131
2	Georgia	10,442	$61.0	$5844
3	Texas	8,844	$84.9	$9595
4	Louisiana	5333	$24.4	$4578
5	Alabama	4508	$28.3	$6280
6	N. Carolina	4226	$28.8	$6810
7	Pennsylvania	3686	$13.2	$3579
8	Tennessee	3397	$24.5	$7212
9	Virginia	3397	$21.0	$6607
10	S. Carolina	3163	$13.7	$4318
Total, top 10		58,671	$455.9	$7771

Source: Insurance Information Institute, State Farm® website http://www.iii.org/fact-statistic/lightning

million, a very different number of claims than calculated by Holle et al. (1996). The average claim reported on this recent website was $7947.

The average cost per claim has generally continued to rise, not only because of inflation but due to the enormous increase in the number and value of consumer electronics. As might be expected from lightning density maps (Chap. 11) as well as population, Florida had the largest number of homeowner insurance claims for lightning losses in 2015, followed by Georgia and Texas (Table 5.2).

Unfortunately, sometimes the loss is much more than electronics. According to a report by the National Fire Protection Association (NFPA), local fire departments in the United States responded to an average of 22,600 residential and nonresidential fires per year started by lightning from 2007 to 2011 (Fig. 5.1). As would be expected, fires started by lightning follow the typical lightning pattern of being more often in the afternoon and during the summer months (Chap. 12).

During this period, an average of 9 deaths, 53 injuries, and $451 million in direct property damage per year were reported (Table 5.3). While the majority of fires

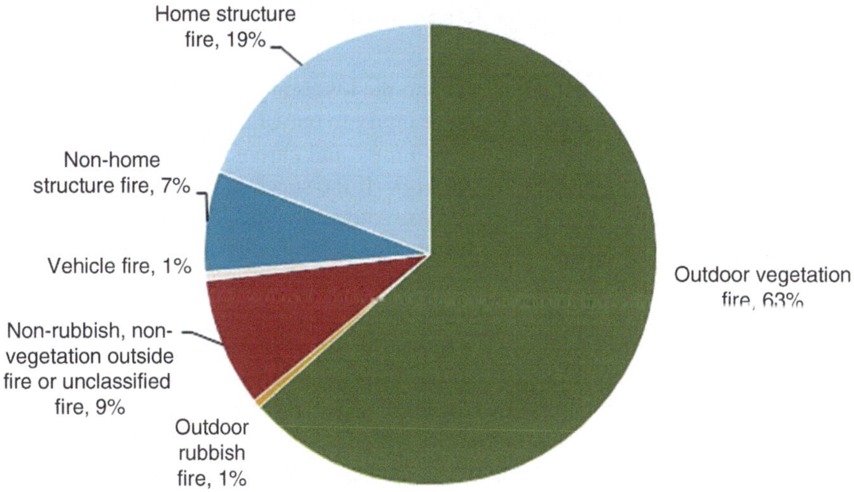

Fig. 5.1 Lightning fires by incident type, 2007 through 2011 (Ahrens 2013). (Reprinted with permission from the report Lightning Fires and Lightning Strikes copyright © 2013, National Fire Protection Association, Quincy, MA. Available through the NFPA Web Site, http://www.nfpa.org/News-and-Research/Fire-statistics-and-reports/Fire-statistics/Fire-causes/Lightning-Fires-and-Lightning-Strikes. All rights reserved)

Table 5.3 Annual average property loss in US dollars from lightning fires to nonresidential properties from 2007 through 2011

Cost	Property type
\$28 million	Storage facilities
\$22 million	Places of assembly, such as houses of worship and restaurants
\$19 million	Nonhome residential properties such as hotels and motels
\$15 million	Mercantile and business properties such as offices, specialty shops and department stores
\$15 million	Industrial and manufacturing facilities
\$3 million	Outside properties
\$3 million	Educational and healthcare facilities
\$3 million	Miscellaneous properties

Source: Insurance Information Institute, http://www.iii.org/fact-statistic/lightning, from NFPA data

were outdoors (73%), those that involved deaths, injuries, and property damage were overwhelmingly associated with home fires even though home fires made up less than one-fifth of the responses to fires (Ahrens 2013). Because only "primary" deaths are counted by the NOAA data collection policy, these "secondary" deaths from lightning-caused fires do not show up in the annual totals of US lightning deaths located at www.lightningsafety.noaa.gov.

5.2 Individual and Family Losses from Injury

Seldom included in any accounting of economic losses are those to the individual or family who has suffered lightning death or injury to one of their family members. In developed countries, survivors with brain injury and chronic pain syndromes may be unable to return to their prior line of work. If the survivor is a child, the parents may not recognize the extent of the injury or know how to access testing and rehabilitation services. It may be difficult to convince schools that the child's deficits are from brain injury and access the additional educational resources that are needed. If it is the primary income producer who is injured, they may not be able to return to work and may need care and advocacy by the uninjured spouse, leading to their loss of work as well. If the family loses their main sources of income, they may lose their mortgage, funds for their children's educations, and health insurance and may face a long battle in court if the injury was work related or due to negligence (Cooper 2001; Cooper et al. 2001; Cooper and Marshburn 2005; Cooper 2005).

In developing countries, all of these also occur, but there may be no clear recognition or understanding of the deficits of the survivor. If the deficits are recognized, there may be no appropriate diagnoses, levels of care, or rehabilitation/retraining available. In some cultures, the injury may be blamed on sin or curses, and the family may need to uproot and move to another community where they will not be ostracized (Cooper 2001; Mulder et al. 2012).

Questions to explore
What losses would your family suffer if you were to be injured or killed by lightning? What if your father or mother were killed or injured and disabled and unable to return to work? Would you be able to continue your studies or live in your current location?

5.3 Personal and Small Business Property Losses: Developing Countries

As might be expected, if US reports vary considerably in quality and completeness, reports from developing countries are nearly nonexistent. Lightning is responsible for losses to individuals and communities, especially in high density, poor infrastructure areas where exposure tends to be high and lightning-safe shelters are largely unavailable. It causes direct damage by killing herds of cattle, pigs, goats, and sheep in areas where personal wealth is measured in animals (Fig. 5.2) and by burning homes and small businesses.

Indirect damage can occur when electrical transmission is interrupted, especially where the electrical grid does not include redundancy or when no alternate sources of energy such as generators are available. Examples of damage include water

a

b

Fig. 5.2 Property damage with deaths of animal. (Top: cattle in South Africa, courtesy of Ian Jandrell. Bottom: goats in Namibia, courtesy New Era Publication Corporation)

pumping in villages and irrigation systems on farms; refrigeration and cooking facilities in restaurants and groceries; data transmission, storage, and credit card processing in small businesses; and a myriad of other impacts on small businesses. These commercial activities and small manufacturing companies may shut down when no lights are available for late night or overnight crews to work. In some areas, repairs may be delayed for considerable periods of time by unavailability of parts and skilled repair teams or by transportation problems to the sites.

Questions to explore
What are the property losses from lightning in your country or region? What are the reporting sources? What methodological flaws, limitations, or biases might these reports have? How would you design a study to collect more accurate statistics? What sources would you use? Is there a minimum period over which you would collect data to prevent annual weather fluctuations from affecting the data?

5.4 Utilities

Power transmission and distribution have a long history of damage from lightning. In most countries, the utility infrastructure is elevated and isolated, often giving poles, lines, towers, and substations significant lightning exposure. While utilities may monitor their power quality, few know the percentage of voltage sags and interruptions that are due to lightning. Losses due to lightning are passed on to the customer since all costs of electricity generation and delivery are part of rate calculations. These are broadly estimated to be in the billions of dollars worldwide. In 2015, over $8 million US dollars were lost due to lightning activity in the Tennessee Valley Authority region alone where it is estimated that 50% of losses are due to lightning (Gamble and Laughlin 2016).

In recent years, wind turbines have become a frequent target of lightning due to their tall shapes and their often isolated locations in open areas and fields. Telephone, cable, and internet wiring mounted on poles is also vulnerable.

The underlying reason for addressing lightning impacts is interruption of power (Cummins et al. 1998; Cummins and Chisolm 2015). In developing countries where electric utility interruption may occur frequently for many non-lightning reasons, any interruption stops delivery of power downstream, leaving many businesses, small and large, with goods that cannot be preserved by refrigeration, communication deficits and often data loss for all of industry, lack of lights to carry on night work, and lack of electricity to power other equipment necessary for running the business, necessitating purchase of generators that often use fossil fuels. Depending on the regulations about emissions for these power sources, they may cause the individual, other people in the home, business, or community to suffer from such impacts as fumes, smoke, carbon monoxide exposure, and risk of fire. Sometimes these outages and losses can accumulate for weeks due to unavailable repair parts or because of remoteness of the damage, particularly when redundancy is not built into the system.

In addition to the indirect impacts of lightning in real time with tangible losses, utilities in developed countries spend a significant amount of their budgets to protect infrastructure, to send repair crews to damaged areas, and to replace equipment when damaged. Power generating stations, substations, and transmission and distribution lines are usually protected (or "hardened") to some extent from lightning.

Such protection can be expensive, especially in locations with higher lightning frequency. As a result, there is always a balance between the high costs of protection versus the costs of lightning-caused interruptions.

Part of the evaluations done by utility providers involve the effect of aging of infrastructure, which may sometimes result in increased susceptibility to lightning-caused interruptions, compared with the cost of replacement or further hardening. These avoidance costs can equal the cost of repairing direct damages from lightning. Separate, exact costs for lightning damage prevention are not generally isolatable, being part of all power grid design, installation, and maintenance. However, if polled, most utility managers would probably acknowledge that the indirect costs in the United States are likely to be in the range of billions of dollars annually.

Oil and gas industry lightning accidents tend to result in major losses with sometimes catastrophic damage and loss of life. It is estimated that lightning accounts for 61% of all accidents in gas storage and processing activities, where natural events are identified as the root cause of the incidents (Necci et al. 2013). A review of fires in the petroleum industry found 150 tank fires in a 52-year period as a result of lightning (Persson and Lönnormark 2004). Protection of such facilities is both expensive and necessary, involving large sums of money globally.

Questions to explore
Are there standard ways of protecting utility structures? Could you develop better or less expensive methods? Are there any ways or data sources that a utility could use to anticipate areas of damage in order to assemble crews for more timely repair? What parts of the infrastructure are most frequently damaged? Could this information allow for stockpiling of these parts to lessen the amount of downtime?

5.5 Industry and Manufacturing

Since society has become massively dependent on systems driven by electric power, outages and voltage sags lasting only a portion of a second can cause major disruptions to computer-based data transfers, manufacturing, and other situations. For example, short lightning-caused interruption can destroy an entire batch of computer chips or other sensitive materials during manufacturing at a very large cost per incident.

5.6 Mining and Agriculture

From the employer's standpoint, mining, agriculture, and other industry, particularly in developing countries, may have to deal with downtime from loss of workers due to injury or death, necessitating funerals and grieving time. Some cultures believe that lightning is God's or a witch's curse on their work or employer, causing workers and their families to flee to avoid further wrath. Mining and manufacturing may also experience downtime or loss of expensive equipment that could have been protected with advance warning forecasts or better hardening of equipment. In developed countries with strong worker compensation and safety laws, long and expensive lawsuits over whether the injured person was affected to the extent they claim can use valuable company resources, not to mention the bitterness it engenders between employees and employer.

From the smallholder's view, death of livestock, often the measure of a family's wealth in developing countries, can be devastating (Fig. 5.2). When a group of cattle or sheep are killed, often while standing under trees, the family's source of income is destroyed because there is usually no compensation from insurance or government agencies. Availability and use of reliable forecasts of lightning, rain, hail, winds, and drought can help farmers to better plan their planting and harvest, increase their income so that they have some leftover to send their children to good schools, and encourage savings as a small monetary reserve to provide for better seed, medical care, and other relatively expensive episodic needs. Without a reserve, many families have to make decisions between survival, the health or education of a child, and debt bondage, which, though illegal, is often the reality in poor countries. In the United States, non-predator cattle losses due to all weather-related causes, including lightning, ranged from a few percent up to 14% in recent years (US Department of Agriculture 2007).

5.7 Banking and Finance

Banking, finance, and data systems can be harmed by very small surges of power, not to mention direct lightning damage. Banking and credit card processing firms are often triple or more hardened to prevent surge-induced errors to accounts. For both large and small businesses, loss of data such as client mailing lists, invoicing, project proposals, and monitoring, as well as other data, may take weeks to rebuild, if it is possible at all. In fact, claims of data loss from lightning have become a large area of insurance fraud in some areas.

5.8 Insurance

A study in the United States (Holle et al. 1996) estimated that 300,000 insurance claims per year are filed, and paid, for lightning damages to homeowners and small commercial customers. The average cost at that time was $916 per claim for a period centered on 1990. Reports since then have shown that while the number of claims has lessened, the cost now exceeds $7000 per paid claim, in part because of the steadily rising presence of lightning-vulnerable electronics in homes (Tables 5.1 and 5.2). These claims alone now account for around a billion dollars per year in the United States, according to numerous web reports in recent years. In addition, insurance claims for businesses, manufacturing, schools, public facilities, and other infrastructure likely are at least of that order of magnitude due to higher costs per claim (Evarts 2010). When such insurance replacement costs are totaled for other developed nations, the costs are certain to be many billions of dollars.

5.9 Forest Fires

In the western United States, nearly half of all forest fires are started by lightning (https://www.nifc.gov/fireInfo/fireInfo_stats_lightng.html). Unfortunately, according to the National Fire Protection Association (NFPA), lightning-induced forest fires burn nearly nine times more acres than human-started fires and will often have multiple starting points. According to NFPA executive reports, the average lightning-caused fire burned 402 acres, compared to the average of 45 acres seen in human-caused wildland fires during the period studied (Ahrens 2013). Similar statistics are found in Canada (https://www.ec.gc.ca/foudre-lightning/default.asp?lang=En&n=48337EAE-1).

Over the 10 years from 2003 to 2012, 42 US firefighters were killed as a result of lightning-caused fires. Four fatalities were at structure fires and 38 at wildland fires, 11 of which were from helicopter crashes (Ahrens 2013). The National Interagency Fire Center tracks lightning fires separately at www.nifc.gov, and forecasts of their occurrence are at www.spc.noaa.gov.

Lightning-caused forest fires occur in many other areas of the world, including Australia, Canada, Russia, and areas where it is often little publicized or apparent to the rest of the world. Direct costs for controlling forest fires can be very large, and damages or destruction of homes and other infrastructure are also expensive. Anywhere that forests overlap or are adjacent to man-made infrastructure has the potential for major damage.

5.10 Aviation

When an aircraft flies at high speed from one location of a thunderstorm to another, differences in charge are started by the plane (Mazur 1989). The most common situation for an aircraft to initiate a lightning flash is during ascent or descent when the plane is changing altitudes and flying through rapid changes in precipitation type at varying temperatures, often in environments where the surface temperature is only somewhat warmer than freezing. It is not unusual for passenger aircraft to be affected by lightning. Although a plane that is struck is able to land safely, the aircraft is required to be inspected for any subtle damage to electronics or structural elements that may compromise future operations before being permitted to fly again. Since poststrike inspections are sometimes very inconvenient and expensive, pilots try to avoid initiating such lightning events. Note that military and smaller civilian aircraft may not have protection from lightning due to the weight of the materials needed to properly surround fuel and electronic components.

Widespread delays can occur at airports due to lightning that affects refueling, catering, baggage handling, passenger movements, and other ground operations. The impact of a single delay at a large airport cascades into downstream effects that affect numerous subsequent flights and passengers at other airports and can result in regional or sometimes national suspensions of operations (Steiner et al. 2013). Methods to minimize airport downtime using lightning data have been explored that attempt to optimize the balance between personnel safety and efficiency of ground operations (Holle et al. 2016).

5.11 Sports and Recreation

In the last two decades or so, prevention measures to avoid lightning that could affect people have grown greatly. These measures have likely been responsible for reducing the number of people killed and injured by lightning in more developed countries (Chap. 6).

Sometimes, the pressure to continue an outdoor event or venue despite the lightning threat is large due to the economic ramifications. At the larger events, loss of television revenues may cause major losses to the stations and networks involved, as well as to people delayed in traveling to games or festivals. The direct loss from one university football game being canceled in the United States can be substantial (http://www.nola.com/lsu/index.ssf/2015/09/lsu_announces_refund_format_fo.html). Similar impacts occur at outdoor concerts, fairs, festivals, and other venues. Fortunately, venue managers have become very aware of the lightning risk, and it has become common for football, soccer, and baseball games, as well as golf tournaments and auto races, to be delayed, suspended, or canceled for lightning. These delays occur at all levels, from professional games, universities, and high schools to local youth leagues (http://www.lightningsafety.noaa.gov/large_venues.shtml).

Personal recreation such as camping, boating, hiking, fishing, and climbing is also prone to being postponed or canceled. It is tempting to continue a long-planned trip despite the lightning risk; otherwise the travel cost of a lost recreation opportunity will become a personal economic loss. Unfortunately, the wrong choice may also result in death or disability for the traveler.

5.12 Economic Development

Per capita gross domestic product (GDP) is directly related to per capita electricity consumption in developing countries (Leung and Meisen 2005). This study indicates a 0.95 correlation between electricity consumption per capita and GDP per capita. Unreliable power sources, due in part to lightning, affect not only individuals and communities but also developing countries where HIV/AIDS, civil unrest, high infant mortality, drought, or poor education are already major impediments to economic growth.

5.13 Historic Sites and Monuments

Valuable monuments and historic sites have been damaged by lightning leading to cultural losses to structures that often represent national identity and history (Fig. 5.3). In developed countries, these sites are often given lightning protection, but this may not be the case in more remote areas or in developing countries.

Fig. 5.3 Lightning damage to a historic temple in Nepal. (Courtesy Shri Ram Sharma)

References

Ahrens M (2013) Lightning fires and lightning strikes. National Fire Protection Association. (http://www.nfpa.org/news-and-research/fire-statistics-and-reports/fire-statistics/fire-causes/lightning-fires-and-lightning-strikes)

Cooper MA (2001) Disability, not death, is the main issue. Nat Weather Dig 25:43

Cooper MA, Marshburn S (2005) Lightning strike and electric shock survivors, international. NeuroRehabilitation 20:43–47

Cooper MA, Marshburn S, Marshburn J (2001) Lightning strike and electric shock, survivors international. Natl Weather Dig 25(1, 2):48–50

Cummins KL, Chisholm WA (2015) Deriving value from lightning data. Transm Distrib World Mag 67:52–56

Cummins KL, Krider EP, Malone MD (1998) The U.S. National Lightning Detection Network™ and applications of cloud-to-ground lightning data by electric power utilities. IEEE Trans Electromagn Compat 40:465–480

Evarts B (2010) Lightning fires and lightning strikes. National Fire Protection Assoc, Fire Analysis Res Div, Quincy, Massachusetts, pp 31

Gamble RD Jr, Laughner T (2016) Lightning studies to improve power quality. In: Preprints of the 6th international lightning meteorology conference, Vaisala, San Diego, California, 18–21 April 2016

Holle RL, Demetriades NWS, Nag A (2016) Objective airport warnings over small areas using NLDN cloud and cloud-to-ground lightning data. Weather Forecast 31:1061–1069

Holle RL, López RE, Arnold LJ et al (1996) Insured lightning-caused property damage in three western states. J Applied Meteor 35:1344–1351

Leung CS, Meisen P (2005) How electricity consumption affects social and economic development by comparing low, medium and high human development countries. Global Energy Network Institute. (http://www.geni.org/globalenergy/issues/global/qualityoflife/HDI-vs-Electricity-Consumption-2005-07-18.pdf)

Mazur V (1989) Triggered lightning strikes to aircraft and natural intracloud discharges. J Geophys Res Atmos 94(D3):3311–3325

Mulder MB, Msalu L, Caro T et al (2012) Remarkable rates of lightning strike mortality in Malawi. PLoS One 7(1):e29281

Necci A, Antonioni G, Krausmann E et al (2013) Accident scenarios caused by lightning impact on atmospheric storage tanks. Chem Eng Trans 32:139–144

Persson H, Lönnormark A (2004) Tank fires: review of fire incidents 1951–2003. Brandsforsk Project 513–021, SP Swedish Natl Testing Res Inst. ISBN 91-7848-987-3

Steiner M, Deierling W, Bass R (2013) Balancing safety and efficiency of airport operations under lightning threats: a look inside the ramp closure decision-making process. J Air Traffic Control 55:16–23

United States Department of Agriculture (2007) Cattle and calves nonpredator death loss in the United States, 2005. USDA-APHIS-VS, CEAH, Natl Animal Health Monitoring System, Fort Collins, Colorado, #N462.0507

Part II
Estimates of People Killed and Injured by Lightning

Chapter 6
Current Global Estimates of Lightning Fatalities and Injuries

Abstract The actual number of worldwide annual deaths and injuries is unknown. The documented number of lightning fatalities around the world currently exceeds 4000 per year and is compiled from national datasets in articles that have been published formally or informally for multiple-year periods ending in 1979 or later. This number is known to be too small, since data are missing in many countries that are expected to have significant vulnerability to lightning in many aspects of society together with a large frequency of lightning flashes. One of the primary goals of this book is to identify how to collect national information from the numerous nations that currently lack data on lightning fatalities and nonfatal injuries in order to update these incomplete estimates.

6.1 Introduction: All Weather Events

Reports of weather-related fatalities worldwide are more likely to concentrate on those that have a large number of casualties. Some natural disaster reporting systems do not include an event unless at least several people are killed in a single instance. While not well documented, it is suspected that heat-related events are globally responsible for more deaths annually than any other weather hazard. The second most common weather impact is due to widespread river flooding during the Asian Monsoon and in many other locations around the world that cause hundreds to thousands of deaths over periods of days or weeks. With respect to individual storms, Cerveny et al. (2017) describe the largest events since 1873 in several categories of natural disasters resulting from storms as summarized in Table 6.1.

Collecting data on incidents that cause only a single or a few deaths and injuries is difficult because they are less likely to be reported in the news, are more likely to "slip through the cracks" of reporting systems, and are infrequently collected in government data programs. Thunderstorms fall in this category. While individual thunderstorms occasionally cause multiple deaths, injuries, and significant damage from their attendant tornadoes, strong winds, hail, lightning, and flash floods, most

M. A. Cooper, R. L. Holle, *Reducing Lightning Injuries Worldwide*,
Springer Natural Hazards, https://doi.org/10.1007/978-3-319-77563-0_6

Table 6.1 Deadliest weather events since 1873

Natural disaster	Location	Number killed	Year
Tropical cyclone/hurricane	Bangladesh	300,000	1970
Tornado	Bangladesh	1300	1970
Hail	India	246	1888
Lightning-indirect deaths	Egypt	469	1994
Lightning-direct deaths	Zimbabwe	21	1975

Cerveny et al. 2017

thunderstorms cause only small numbers of deaths, injuries, and damage, and most are isolated to a specific locality for a time of only a few hours, instead of the multiday, large-scale impacts such as heat, hurricanes, and flooding.

6.2 Lightning Events

The United States National Weather Service defines severe weather from thunderstorms as tornadoes, strong winds over 58 miles/hour (93 km/h), and hail at least 1 inch (2.5 cm) in diameter (www.spc.noaa.gov). Although lightning occurs much more frequently in area and time than these localized phenomena, it is not defined as severe weather. A lightning event can cause one or a few deaths and injuries at a time, but a very large number of such cases are spread over the globe throughout the year. In the developed portions of the world, most lightning events kill or injure one person at a time (Curran et al. 2000). This lower toll per event is attributable to the widespread availability of safe areas such as substantial buildings and fully enclosed all-metal vehicles, as well as lightning injury prevention education, improved medical care, lesser lightning density in temperate zones where many developed countries are located, and other factors that are identified throughout this book.

In developing regions, ten or more people are sometimes reported as killed in a single lightning event. This is due to a combination of:

- Increased exposure due to labor-intensive work practices, including agriculture.
- Lack of lightning-safe structures and vehicles.
- Different outcomes where multiple people are killed in lightning-unsafe homes, open churches, and school rooms rather than injured as occur in more developed countries (Chap. 7).
- Large lightning density in some regions of the tropics and subtropics.
- Individual incidents are unreported or infrequently reported compared with events that have a large number of casualties, leading to skewing of the reports toward instances with larger numbers of casualties.
- Inadequate or incomplete collection of the number of lightning victims in these countries (Chap. 6).

As a result, the summation of sporadic but frequent global lightning fatalities accumulates to become very large totals. But they are especially elusive to count, in part due to the smaller number of deaths and injuries per instance from lightning that are often unreported, compared with the massive weather-related events. Occasionally, the media will group several separate lightning events with fatalities in a nation or region that occur during a period of a few days. This grouping leads to the realization of how often such sporadic, but multiple, fatalities are actually occurring. However, the actual number may well be very much underreported. A recent event in Bangladesh (Holle and Islam 2016) indicated that 64 people were reported as killed by lightning during the course of a 2-day period. Similar multiple-fatality groupings of lightning fatalities have been reported in Uganda and India (Chap. 6).

6.3 Collection of Lightning Fatality and Injury Data

Fatalities are the most reliable dataset to consider, as they are more likely to be documented by medical or death records or to be reported by the news media. Unfortunately, data on nonfatal injuries are often much harder to collect. Many individuals do not seek medical care at the time of the incident. When a visit to a medical facility does occur, the visit diagnosis may not be coded as lightning-caused in medical or other datasets. Additionally, in some developing countries, lightning injury is considered shameful or as a retribution for sins so that survivors are reluctant to report them or to seek care.

While the relatively common single-person fatality is not certain to reach a database, it is much more likely to be reported than an isolated single injury. As a result, fatality numbers are more reliable and have become the preferred metric to use when comparing and collecting data across the globe.

In developed nations, the normal ratio of fatalities to nonfatal injuries is one fatality out of every ten injured when full medical reporting is examined, as discussed in Chap. 15 (Cherington et al. 1999). Unfortunately, the ratio of ten injuries per death is unlikely to apply to lesser-developed countries where deaths, let alone injuries, go unreported and where more people are likely to be affected such as in a single event due to labor-intensive agriculture and lack of lightning-safe dwellings and vehicles (Holle 2010, 2016b).

No database exists of global lightning deaths, injuries, or damages (Chap. 5). The primary reason is the widespread but mainly isolated impacts of lightning in time and space. Data in some more developed countries have been compiled in recent years, but many nations do not keep or have any such records.

There are several methods to collect weather-related casualty data. Examples of the types of data collection methods are identified in the right column of Table 6.2:

- *A, Meteorological*: Lightning summaries are sometimes developed by national meteorological agencies as part of the monitoring of all types of weather impacts in their country, including heat, cold, tropical cyclones, and other phenomena. The quality often varies, but this is at least a national approach.

Table 6.2 Published annual lightning fatality rates per million people and number of fatalities by country with a dataset with its last year being 1979 or later. Data collection type A, national meteorological agency; B, medical records; C, personal data collection from variety of sources; D, print media; E, natural hazards database; F, mixture of types

Continent Country	Annual fatality rate per million	Fatalities per year	Data collection type
Africa			
Burundi	2.5	26	C
Malawi	84.0	1008	F
South Africa	6.3	264	B
Swaziland	15.5	15	F
Uganda	0.9	30	F
Zimbabwe	14 to 21	100 to 150	F
Asia			
Bangladesh	1.6	251	F
China	0.3	360	E
India	2.0	1755	E
Japan	>0	2	C
Malaysia	0.8	22	F
Mongolia	1.5	5	A
Singapore	1.5	3	F
Sri Lanka	2.6	49	C
Australia			
Australia	0.1	2	D
Europe			
Austria	>0	1	D
France	0.2	11	B
Greece	0.1	5	F
Lithuania	0.1	2	A
Poland	0.3	8	F
Switzerland	0.4	2	E
Turkey	0.4	28	F
United Kingdom	>0	2	F
North America			
Canada	0.2	9	A
Mexico	2.7	230	B
United States	0.1	31	A
South America			
Brazil	0.8	132	F
Colombia	1.8	76	B

Updated from Holle 2016a, c

- *B, Medical*: Medical reporting systems that use death certificates may be very reliable in countries with a uniform national system (Navarrete-Aldana et al. 2014). However, in developing countries, the generation of birth and death certificates may be initiated only for hospital cases or other medical or official settings, leaving as many as two-thirds of the common births and deaths undocumented (Richard Tushemereirwe, Uganda, personal communication).

- *C, Personal*: Sometimes, an individual will personally undertake the collection of all lightning casualties in a nation by making a survey of newspapers such as that done in the 1890s by Kretzer in the United States and summarized by Holle et al. (2005). A less common approach is to make personal interviews of as many people as possible as reported for Malawi by Mulder et al. (2012).
- *D, Media*: In nations with well-developed national media coverage over many years, newspapers and, more recently, their online versions may capture nearly all incidents, although details of the injuries may or may not be accurate from these sources.
- *E, Natural hazards*: Sometimes the natural hazard community within a nation has compiled an exhaustive database, such as in Switzerland (Badoux et al. 2016). However, there is a tendency in such collection systems to omit events where few people were affected such as is often the case with lightning. As a result, cases of individual or small groups of injuries and deaths from whatever cause may be overlooked, even in these collections.
- *F, Mixture*: Finally, it may be necessary to use a combination of all available data sources such as the approach for Bangladesh used by Dewan et al. (2017). Great care must be taken to cross-reference each case to avoid duplications.

Questions to explore

What is the most important factor influencing the collection of lightning fatality and nonfatal injury data in your country? Who is most likely to have access to the datasets? How complete are the data across the country, and how long are the periods when they have been available?

A recent example of the challenges in lightning casualty collection is a report from India. Illiyas et al. (2014) reported an annual average of 1755 fatalities. In contrast, Singh and Singh (2015) only found 159 annual fatalities in India, which, in context, appears to be much too small. Nevertheless, the difference indicates the difficulty of collecting lightning fatality totals on a national scale.

Regardless of the collection method, datasets may be dominated by more newsworthy events that are relatively rare but result in the larger number of deaths and injuries than the single death or injury. Because of the widely dispersed but frequent impacts of lightning, ideally, a data collection system should be established that covers the entire country and is able to capture the single-person events as well as the more spectacular multiple injury events in each country. At least two recent studies, one by Thacker et al. (2008) for the United States and another by Badoux et al. (2016) for Switzerland, have collected natural hazard data for incidents involving only a single lightning casualty. Unfortunately, this inclusion of the frequent single lightning events occurs infrequently and irregularly.

Another approach is to estimate global fatalities with lightning and population data, rather than direct collection method for fatalities. This approach (Roeder et al. 2015) multiplied population times lightning density in the United States to make a reasonably good estimate of lightning fatalities. The next step will be to verify and refine this method by applying it to a lesser-developed country that has excellent death records, such as Colombia (Navarrete-Aldana et al. 2014). If this approach is successful, the method may be able to be used as a tool to make a reasonable estimate of the lightning fatality risk using available population distribution data and global lightning occurrence data. Such an approach is feasible in the future but involves sophisticated manipulation of large datasets comprised of population, lightning, and global information system information.

6.4 Estimates of National Fatality Rates by Continent

Three estimates of the global lightning fatality totals have been made in the last several years. The lower estimate is several thousand fatalities per year by Gomes and ab Kadir (2011). A middle value is 6000 per year by Cardoso et al. (2014). The largest is 24,000 lightning fatalities per year by Holle and López (2003) extrapolating from known conditions and fatalities for a temperate country. Extrapolations may be much too low for tropical and subtropical regions that often have frequent lightning activity and where people may have continual exposure to lightning due to a lack of available safe housing and during daytime labor-intensive work activities such as agriculture.

Lightning fatality data from 28 countries (Table 6.2 and Fig. 6.1) have been published in the last quarter century in formal and informal papers (updated from Holle 2016a,c). When the numbers are added for these 28 countries, an annual total of 4429 fatalities is obtained (Holle 2016c). The largest published national fatality totals are 1755 per year in India and 1008 per year in Malawi, although it is not always well documented how these figures were gathered. The total of 4429 fatalities is an underestimate since there are more nations without published data than we have with published data (Table 6.3). The countries of most interest with high fatality rates, but without published summaries, are in lesser-developed regions, many of them with a large amount of lightning.

The following are comments on each continent in Fig. 6.1 and Table 6.2. The references at the end of this chapter indicate the data sources. The following discussion begins in the upper left of Figure 6.1:

- *North America*: Canada and the United States have very low fatality rates (< 0.5 deaths/million/year) in recent years, while Mexico has an intermediate range (0.6 to 5.0 deaths/million/year).
- *South America*: Analyses of Brazil and Colombia data show intermediate rates.
- *Europe*: All eight nations with data in recent years show low rates. However, there are many nations without data.

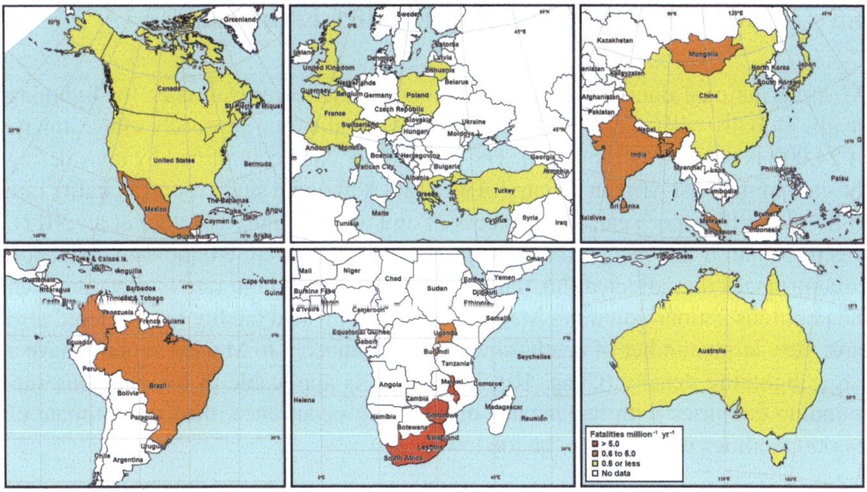

Fig. 6.1 National lightning fatality rates per million people per year by continent. Red shading indicates rates >5.0 fatalities per million per year, orange is 0.6 to 5.0, and yellow is 0.5 or less. White indicates no national summaries have been published for periods ending in 1979 or later. (Updated from Holle 2016a, c)

Table 6.3 Current estimates of worldwide annual lightning fatalities

Number of annual deaths	Publication
Several thousands	Gomes and ab Kadir (2011)
6000	Cardoso et al. (2014)
24,000	Holle and López (2003)
4429	Updated from Holle (2016c)

- *Africa*: Malawi, South Africa, and Swaziland all have documented large lightning fatality rates (> 5 deaths/million/year), while Burundi and Uganda have moderate rates. It is very likely that many other countries have large rates, but the number of nations without fatality data needs to be increased as soon as possible in order to assess what lightning safety actions are to be taken.
- *Asia*: China and Japan are reported to have low fatality rates, while Bangladesh, India, Malaysia, and Mongolia have intermediate rates. Several Southeast Asian countries where high lightning fatality rates would be expected from risk factors and lightning density have had no published multi-year studies to date.
- *Australia*: The rate is very low in this developed nation.

6.5 Status of Global Lightning Fatality Estimation

As more national datasets are being published, the lower estimates for lightning deaths globally will quickly be surpassed. How far the total approaches the estimate of 24,000 is unknown (Table 6.3). The uncertainty can be amplified by considering the case of the East African country of Malawi. A national study found a fatality rate of 84 per million per year, far in excess of any other country (Mulder et al. 2012). It is uncertain if this is a correct assessment or if there were unique data collection circumstances that affected this total (Table 6.4). If this high rate is accurate, then the populous nations adjoining Malawi without published fatality totals should also have very large number of deaths since regions adjacent to Malawi actually have a larger lightning density (Chap. 10). If that rate is applicable to the populous surrounding countries with larger rates of lightning occurrence, then the estimate of 24,000 fatalities per year may be too low.

Table 6.4 Factors that can change the estimate of 24,000 worldwide lightning fatalities per year

Factor	Change	Change in fatalities
Area of very frequent lightning	Too small	Increase
	Too large	Decrease
Fatality rate of six deaths per million people	Too low	Increase
	Too high	Decrease
Rural agricultural settings in frequent lightning areas compared to the United States and Western Europe in 1900	More rural Less rural	Increase Decrease
Buildings occupied by people in frequent lightning areas compared to the United States and Western Europe in 1900	Less substantial More substantial	Increase Decrease
Organized recreational sports compared to the United States and Western Europe in 1900	More	Increase
Meteorological forecasts and warnings	Improved	Decrease
Awareness of the lightning threat through education, planning, and detection	Enhanced	Decrease
Medical care and emergency communications	Enhanced	Decrease
Other socioeconomic changes	Unknown	Unknown

Adapted from Holle and López 2003

References

Badoux A, Andres N, Techel F et al (2016) Natural hazard fatalities in Switzerland from 1946 to 2015. Nat Hazards Earth Syst Sci 16:2747–2768

Cardoso I, Pinto O Jr, Pinto IRCA et al (2014) Lightning casualty demographics in Brazil and their implications for safety rules. Atmos Res 135-136:374–379

Cerveny R, Bessemoulin P, Burt CC et al (2017) WMO assessment of weather and climate mortality extremes: lightning, tropical cyclones, tornadoes, and hail. Weather Clim Soc 9:487–497

Cherington M, Walker J, Boyson M et al (1999) Closing the gap on the actual numbers of lightning casualties and deaths. In: Preprints of the 11th conference on applied climatology, American Meteorological Society, Dallas, 10–15 January 1999

Curran EB, Holle RL, López RE (2000) Lightning casualties and damages in the United States from 1959 to 1994. J Clim 13:3448–3453

Dewan A, Hossain MF, Rahman MM et al (2017) Recent lightning-related fatalities and injuries in Bangladesh. Weather Clim Soc 9:575–589

Gomes R, ab Kadir MZA (2011) A theoretical approach to estimate the annual lightning hazards on human beings. Atmos Res 101:719–725

Holle RL (2010) Lightning-caused casualties in and near dwellings and other buildings. In: Preprints of the 3rd international lightning meteorology conference, Vaisala, Orlando, 21–22 April 2010

Holle RL (2016a) A summary of recent national-scale lightning fatality studies. Weather Clim Soc 8:35–42

Holle RL (2016b) Lightning-caused deaths and injuries related to agriculture. In: Preprints of the 33rd international conference on lightning protection, Estoril, 25–30 September 2016

Holle RL (2016c) The number of documented global lightning fatalities. In: Preprints of the 33rd international conference on lightning protection, Estoril, 25–30 September 2016

Holle RL, Islam AKMS (2016) Summary of multiple-fatality lightning events during mid-may 2016 in Bangladesh. In: Preprints of the 33rd international conference on lightning protection, Estoril, 25–30 September 2016

Holle RL, López RE (2003) A comparison of current lightning death rates in the U.S. with other locations and times. In: Preprints of the international conference on lightning and static electricity, Royal Aeronautical Society, Blackpool, 16–19 September 2003

Holle RL, López RE, Navarro BC (2005) Deaths, injuries, and damages from lightning in the United States in the 1890s in comparison with the 1990s. J Appl Meteorol 44:1563–1573

Illiyas FT, Mohan K, Mani SK et al (2014) Lightning risk in India: challenges in disaster compensation. Econ Polit Wkly XLIX:23–27

Mulder MB, Msalu L, Caro T et al (2012) Remarkable rates of lightning strike mortality in Malawi. PLoS One 7(1):e29281. 09 January 2012

Navarrete-Aldana N, Cooper MA, Holle RL (2014) Lightning fatalities in Colombia from 2000 to 2009. Nat Hazards 74:1349–1362

Roeder WP, Cummins BH, Cummins KL et al (2015) Lightning fatality risk map of the contiguous United States. Nat Hazards 79:1681–1692

Singh O, Singh J (2015) Lightning fatalities over India: 1979-2011. Meteorol Appl 22:770–778

Thacker MTF, Lee R, Sabogal RI et al (2008) Overview of deaths associated with natural events, United States, 19792004. Disasters 32:303–325

Chapter 7
Lightning Fatalities Since 1800

Abstract Published lightning fatality data in terms of the rate per million people per year show similarities between (a) developed countries before the early twentieth century and (b) current lesser-developed nations. It is apparent that the reduction in manual, labor-intensive agriculture and the increasingly widespread availability of lightning-safe dwellings and vehicles since the early 1900s have been major factors leading to a huge reduction in the fatality rates in developed countries. Unfortunately, that has not yet taken place in lesser-developed countries and regions of the globe. In the developed countries of the United States, Canada, Western Europe, Japan, and Australia, the lightning fatality rate has now reached a level of less than 0.2 per million people per year. However, the rates continue to be very high in developing countries as evidenced by recent data that span the transition from the late twentieth into the twenty-first century. Recent examples of very high fatality totals in Southeast Asia and Africa are also described.

7.1 Nineteenth-Century Fatalities

Lightning fatality data from as early as the nineteenth century were summarized for a few countries by Holle (2008). Figure 7.1 shows population-weighted fatality rates in this summary for seven regions of Europe and Australia that have data from the decade starting in 1810. Note that some decadal rates exceed three fatalities per million per year, which is a much greater rate than the current European rates of less than one fatality per million per year. While the fatality data are often incomplete during the 1800s in most countries, the large fatality rates for these early years are indicative of a very different socioeconomic situation compared with the present time in these regions.

Many of these countries have relatively low flash densities (Figs. 11.2 and 11.3), so the population-weighted fatality rates are all the more exceptional compared with the present rates. Although many of these regions were relatively well developed by global standards at the time, lightning protection was not widely applied in practice. During the day, most agriculture was manual and labor-intensive, not mechanized as it is now. During the day and evening, workplaces, dwellings, schools, and other buildings were not safe from lightning. There were no lightning-safe, fully enclosed, metal-topped vehicles during the nineteenth century, and grounded plumbing and

M. A. Cooper, R. L. Holle, *Reducing Lightning Injuries Worldwide*,
Springer Natural Hazards, https://doi.org/10.1007/978-3-319-77563-0_7

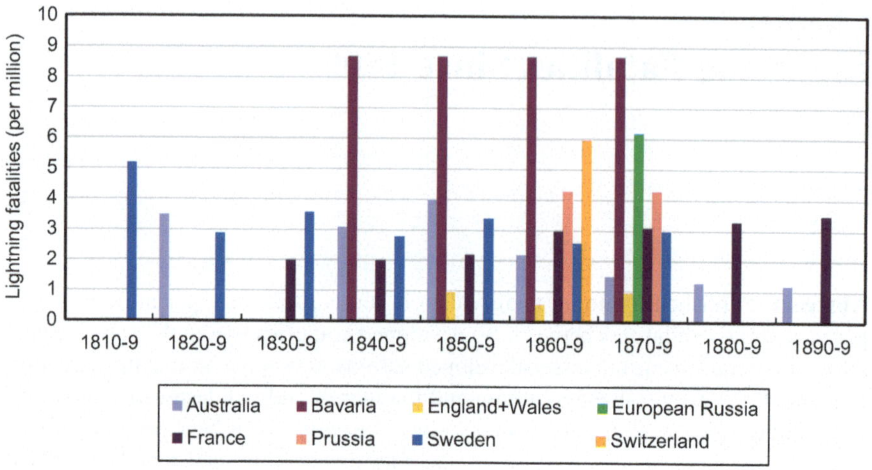

Fig. 7.1 Lightning deaths per million people per year for Australia and seven regions in Europe by decade during the nineteenth century. (Holle 2008)

wiring had not been widely introduced except in some locations of the most developed cities. As a result, the vulnerability of the population was similar to the present situation in lesser-developed regions of the world such as much of rural Africa, South America, and Southeast Asia.

For these reasons, Fig. 7.1 provides a useful background for the baseline population-weighted lightning fatality rate in a society that was very exposed to the threat of lightning at all times prior to the widespread reduction in labor-intensive agriculture and the introduction of lightning-safe buildings and vehicles. The population-weighted rate of lightning fatalities was often as high as three per million per year in these regions that do not have especially high lightning densities, so the influence of socioeconomic changes in Europe and Australia during the last two centuries that resulted in plummeting fatality rates is all the more evident.

Which countries had the highest and lowest fatality rates during the nineteenth century? What region had the first documented decadal lightning fatality rate?

7.2 Twentieth-Century Fatalities

Lightning fatality data have been published for a different combination of eight European countries during the twentieth century – note that only France and England and Wales are repeated from the nineteenth century (Holle 2008). Figure 7.2 shows a rapid reduction in population-weighted fatality rates in these nations from the previous century. The scale has remained at a maximum of ten fatalities per million

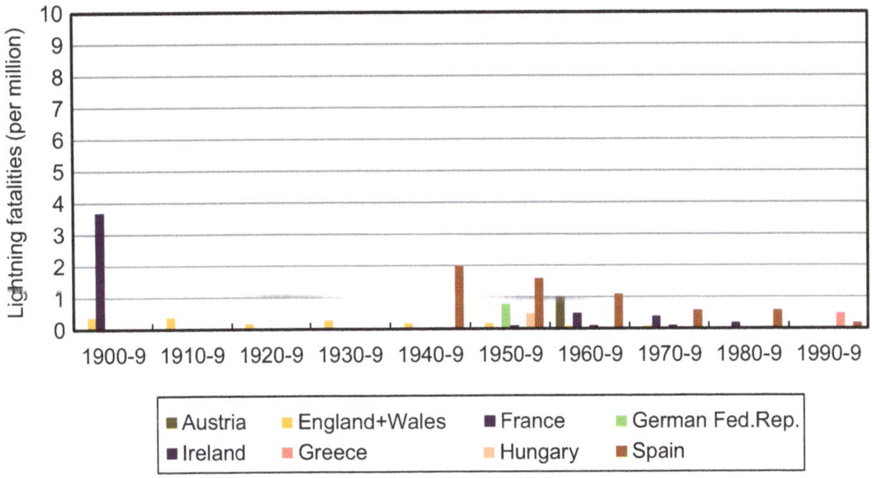

Fig. 7.2 Lightning deaths per million people per year for eight countries in Europe by decade during the twentieth century. (Holle 2008)

per year in order to facilitate comparisons between centuries. Several factors are identifiable; these are not necessarily in order of importance since such socioeconomic ratings can be arbitrary. The most important factor may be the introduction of grounded wiring and plumbing in buildings. Additionally, there is a massive reduction in the amount of time spent by people in manual, labor-intensive outdoor agriculture due to mechanization in the developed regions where many activities now take place when workers are inside fully enclosed metal-topped vehicles. For example, it took much less time to plow or harvest a large field in the late twentieth century compared with the nineteenth century. Another sometimes-overlooked factor is the introduction of fully enclosed, metal-topped vehicles that are now within a short distance of most people living and working in all aspects of their daily activities. In addition, to be added to the list of advancements leading to a reduction in lightning casualties are lightning safety education, medical advances, and lightning monitoring in real time during the twentieth century (Cooper and Holle 2012).

A global view outside of Europe during the twentieth century is shown in Fig. 7.3. Note the United States trend, starting from nearly five fatalities per million people per year in the first decade of the twentieth century, which decreased steadily to a very much small rate at the end of the century. Similar trends are observable for the other three more developed countries of Australia, Canada, and Japan. However, data from Zimbabwe and South Africa that have become available in the last two decades show them to belong to a different sample. These large population-weighted fatality rates resemble those in Europe and Australia during the nineteenth century. It is reasonable to conclude that many of the same factors are applicable to nineteenth century Europe and late twentieth century South African nations. Those factors are lightning-unsafe buildings, manual, labor-intensive agriculture, and a lack of safe vehicles, education, medical treatment, and lightning awareness, all leading to 24/7 vulnerability for entire families, not only outdoor workers.

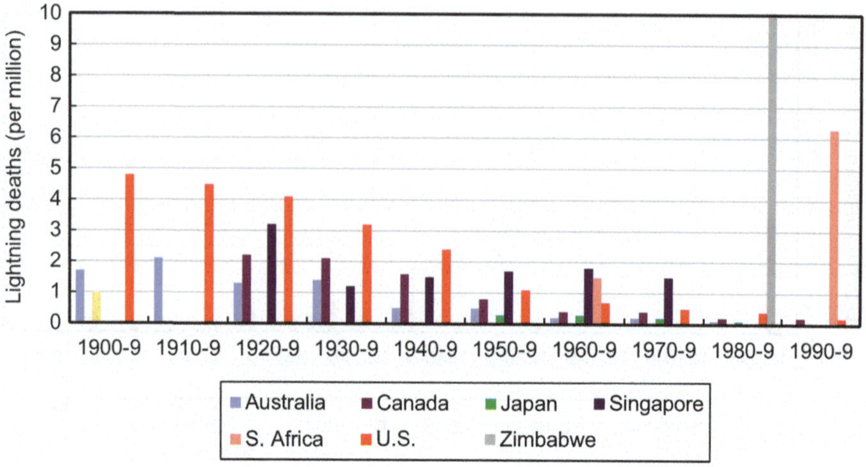

Fig. 7.3 Lightning deaths per million people per year for eight countries outside Europe by decade during twentieth century. The 1990's Zimbabwe rate is 17.8. (Holle 2008)

Questions to explore
Which countries had the highest and lowest fatality rates during the twentieth century outside of Europe? Which nations had the largest changes during the course of the twentieth century?

7.3 US Fatality Rates in the Twenty-First Century

For the United States, Fig. 7.4 shows data from the start of the twentieth century to the present time. The fatality rate exceeded six per million per year during a year just after 1900 but has dropped to near or below 0.1 in the latest years. This rate decrease of more than an order of magnitude has no parallel in any other weather-related phenomenon in the United States. It is difficult to quantify some of the factors that can be expected to contribute to this downward-trending time series; however, some comments are as follows:

- *Rural and agriculture shifts*: Figure 7.4 also shows the percent rural population in the United States to have decreased from 60% in 1900 to less than 20% at the present time (López and Holle 1998). Since the population of the country has more than tripled since 1900, the absolute number of people living in census-designated rural areas has stayed about the same. Rural percentage is important since it is a representation of one of the causes of the reduction in the lightning fatality rate during this period. Manual labor-intensive agriculture was a significant component of US occupations in 1900 when more than half of the people lived in rural settings. However, agricultural mechanization has occurred such

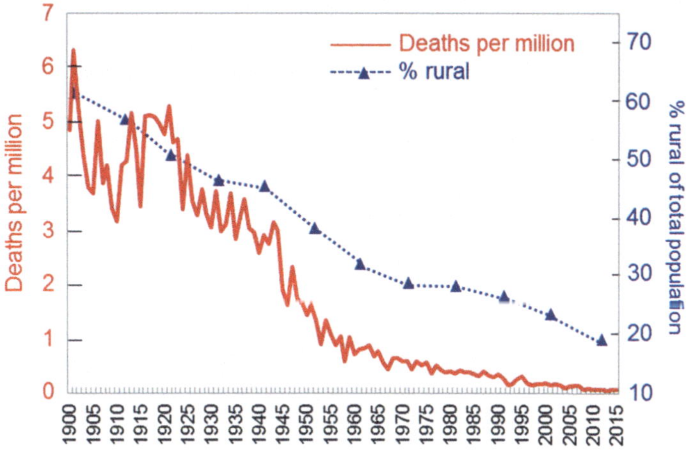

Fig. 7.4 Annual time series for the United States from 1900 to 2015 of lightning deaths per million people shown by solid red line and percent rural population by dashed blue line. (Updated from López and Holle 1998)

that the proportion of time spent by workers outside exposed to lightning during the course of farming is greatly reduced.

- *Vehicles*: Fully enclosed metal-topped vehicles are now nearly universally available in the United States. Figure 7.5 shows how the number of vehicles per population has a similar trend to the rural time series, with vehicles shown in an inverted scale. Notice that there is now nearly one vehicle per person in the United States. Many other technological and societal changes could be mentioned with similar trends since 1900, but the accessible, mobile lightning safety provided by vehicles appears to be important.
- *Buildings*: The percentage and number of grounded lightning-safe buildings have greatly increased since 1900 to where nearly all workplaces and dwelling are now lightning-safe inside (Holle 2010). In recent years, the only fatalities inside dwellings in the United States are due to the elderly, very young, or mentally or physically disabled being unable to escape nighttime lightning-caused fires. All other lightning-caused fatalities related to structures occur to people outside a dwelling or building such as in the yard, under a tree in the yard, mowing the lawn, and other locations not surrounded by a substantial structure. Injuries inside buildings are most often related to wiring and plumbing where a strike to a structure can be expected to be dangerous if a person is in contact with these paths.
- *Medical understanding*: A large improvement has occurred, mainly in the last two decades such that some of what would have been fatally injured people in the past now survive (Cooper et al. 2017). The improvements into the medical insight of lightning casualties are included in Chaps. 3 and 4 of this book.
- *Lightning safety education and outreach*: Rules and approaches for lightning safety have been rewritten since the early 1990s (Jensenius 2016). The emphasis

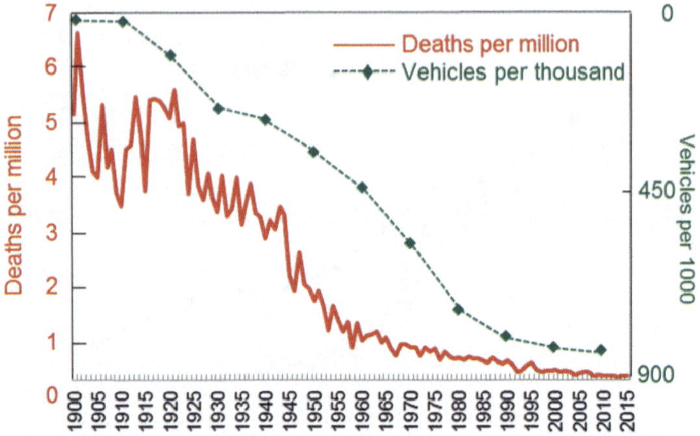

Fig. 7.5 Annual time series for the United States from 1900 to 2015 of lightning deaths per million people shown by red solid line and number of vehicles per thousand people shown by green dashed line. (Note that vehicle scale is inverted)

is now on the two lightning-safe locations of properly grounded buildings and metal-topped fully enclosed vehicles described throughout this book. Prior advice had emphasized speculative measures of limited or no value concerning posture, location, and other features and were most often related to the direct strike that is now considered to be an infrequent source of lightning casualties (Chap. 2).

- *Real-time lightning observations*: Lightning occurrence data have been available for several decades (Cummins and Murphy 2009). Lightning tracking is now available through all types of media such that its awareness has completely changed the perception from that of lightning being vague and random to a phenomenon that is a now a much better-known quantity.

The trends shown for the United States in Fig. 7.4 indicate a precipitous drop in lightning fatality rates to below 0.1 in the latest years. For Canada (Fig. 7.6), a very similar trend is evident since data began to be collected in the late 1930s (Holle and López 2003) that is likely due to the same socioeconomic factors listed above for the United States, although the rates are still lower than in the United States due to the smaller lightning density in Canada. In contrast, Fig. 7.6 shows that Spain had a large lightning fatality rate per million people during the 1940s and 1950s until there was a major population shift to urban regions, and then the fatality rate reached values similar to the United States and Canada (Holle and López 2003).

Questions to explore

Is there any way to determine which is more important, rural or vehicle changes, in the United States during the twentieth century? What other factors can be measured related to the changes in lightning fatality rate during the twentieth century in the United States?

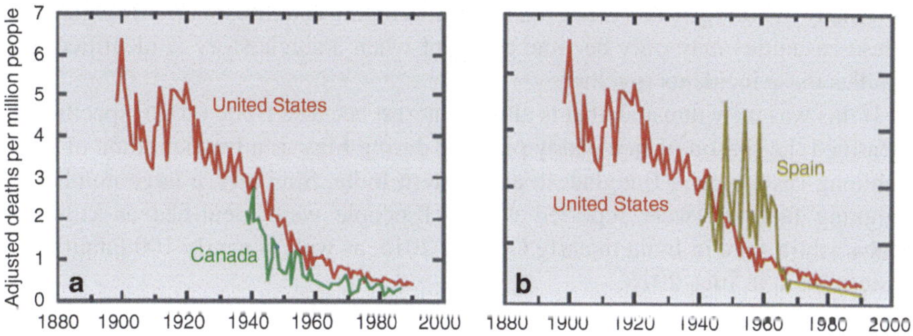

Fig. 7.6 Time series of yearly lightning deaths normalized by population for United States compared to Canada (**a**) and Spain (**b**). (Holle and López 2003)

7.4 Fatality Rates in the Twenty-First Century Outside the United States

Lightning fatality rates from the twenty-first century are included in Fig. 6.1 and Table 6.2 for 28 published national studies beginning in 1979 or later. Lightning fatality rates across the globe at the present time have a very wide range. At one end of the scale are the very low fatality rates that have drastically decreased since a century ago in developed countries such as the United States. These low fatality rates and sharply decreasing trends over the last century also apply to Western Europe, Canada, Japan, and Australia as shown in Fig. 7.3. In contrast, many regions continue to have large lightning fatality rates that correspond with those found in the developed countries a century ago. Figures 7.4 and 7.5 show two of the factors in this change, that of a shift from rural to urban setting resulting in less agriculture participation, and the availability of lightning-safe vehicles.

7.5 Recent Major Fatality Events

In Bangladesh, during mid-May 2016, multiple reports arrived through websites that a large number of people were killed by lightning. The sequence of reports is of interest. First, it was reported that 30 deaths had occurred; in subsequent hours, the total kept rising until 81 was reached within a few days (Holle and Islam 2016). It is apparent that reporters and correspondents in a wide variety of locations within Bangladesh contributed their local news reports to a national collection that most likely would not have reached the media if only single-fatality incidents had been reported on a local basis.

Were these unusually large numbers a result of a focused recognition of the problem, or were they truly unusual? If these are indeed previously unreported but common events, then the data collection system has been significantly underestimating how many lightning deaths are occurring in developing countries such as Bangladesh. These casualties may only become apparent when a single story is identified that bundles these incidents together.

If this was truly unusual, that is also of interest because Holle (2016) specifically identified the season of rice paddy planting during May as a frequent time of prior lightning casualties in Bangladesh and eastern India. Similarly, a large number of lightning fatalities were reported when 41 people were identified as killed in Maharashtra state in India in early October 2015, as well as nearly 100 fatalities in eastern India in June 2016.

Additional bundling of separate events has occurred in several African countries with similarly large numbers of fatalities that spurred local attention. As all three of these stories developed, it appears that reporters in various regions of the affected countries started looking for lightning fatalities and sent them to comprise a single collection. From this process, once a focused search takes place, numerous incidents are identified that may not otherwise have been known. Reports of these multiple-casualty events can be used by the media to raise the level of lightning awareness in their country. In fact, the media can contribute to public awareness of the risk of lightning injury by making them widely available in popular media.

References

Cooper MA, Holle RL (2012) Lightning safety campaigns – USA experience. In: Preprints of the 31st international conference on lightning protection, Vienna, 2–7 Sept 2012

Cooper MA, Andrews CJ, Holle RL et al (2017) Lightning-related injuries and safety. In: Auerbach (ed) Wilderness Medicine, 7th edn. Elsevier, Philadelphia, pp 71–117

Cummins KL, Murphy MJ (2009) An overview of lightning locating systems: history, techniques, and data uses, with an in-depth look at the U.S. NLDN. IEEE Trans Electromagn Compat 51:499–518

Holle RL (2008) Annual rates of lightning fatalities by country. In: Preprints of the 2nd international lightning meteorology conference, Vaisala, Tucson, 21–23 Apr 2008

Holle RL (2010) Lightning-caused casualties in and near dwellings and other buildings. In: Preprints of the 3rd international lightning meteorology conference, Vaisala, Orlando, 21–22 Apr 2010

Holle RL (2016) Lightning-caused deaths and injuries related to agriculture. In: Preprints of the 33rd international conference on lightning protection, Estoril, 25–30 Sept 2016

Holle RL, López RE (2003) A comparison of current lightning death rates in the U.S. with other locations and times. In: Preprints of the international conference on lightning and static electricity, Royal Aeronautical Society, Blackpool, 16–18 Sept 2003

Holle RL, Islam AKMS (2016) Summary of multiple-fatality lightning events during mid-May 2016 in Bangladesh. In: Preprints of the 33rd international conference on lightning protection, Estoril, 25–30 Sept 2016

Jensenius J (2016) NOAA's lightning safety awareness efforts – what we've accomplished in 15 years. In: Preprints of the 6th international lightning meteorology conference, Vaisala, San Diego, 18–21 Apr 2016

López RE, Holle RL (1998) Changes in the number of lightning deaths in the United States during the twentieth century. J Clim 11:2070–2077

Chapter 8
Locations and Activities of Lightning Casualties

Abstract In developed regions of the globe, lightning protection is provided by the ready availability of lightning-safe structures and fully enclosed metal-topped vehicles during most daily activities of most people. In developing economies, in contrast, the lack of such lightning-safe structures and vehicles makes lightning safety very difficult to achieve. At night, people are inside dwellings that provide little or no lightning protection due to their inadequate construction. During the day, workers in agricultural fields and many other workplaces, as well as students at schools, rarely have a safe place to avoid the lightning threat. The primary solution is to install lightning protection on buildings in the fields and on dwellings and school buildings. Field workers and students then need to be instructed to go inside the safe locations when lightning occurs.

8.1 Developed Nations

The enormous reduction of lightning fatalities in developed countries has been described in Chap. 7. The rate has decreased in the United States from as high as six annual fatalities per million people in the early 1900s to 0.1 annual fatalities per million people in recent years (Fig. 7.4). This rate decrease is also apparent in Canada (Mills et al. 2010), Western Europe (Gourbière 1998; Elsom 2015), Japan, and Australia (Coates et al. 1993). In these regions, people live and work in lightning-safe structures and typically have vehicles nearby that provide protection.

The rural percentage of a nation's population is a readily accessible statistic that is updated by decade for every region and nation of the world. It serves as a proxy for the percentage of the population engaged in agricultural activity that has become highly mechanized in developed nations. The proportion of people living in census-designated rural settings in the United States has decreased by two-thirds from the 60% rural proportion 100 years ago to the current value of under 20% rural (Fig. 7.4). The low rural percentage in developed nations indicates that relatively few people are being exposed to lightning during the occupation of farming compared with a century ago (Holle 2016b).

M. A. Cooper, R. L. Holle, *Reducing Lightning Injuries Worldwide*,
Springer Natural Hazards, https://doi.org/10.1007/978-3-319-77563-0_8

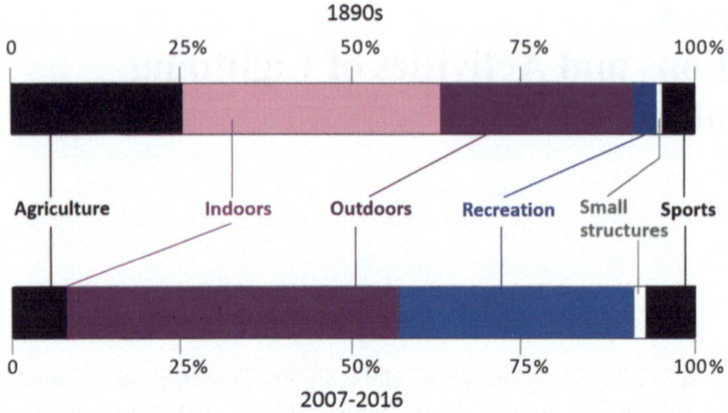

Fig. 8.1 Comparison of the percentage of types of US lightning fatalities in the 1890s versus 2006 through 2015. (Updated from Holle et al. 2005 and Holle 2016a)

The following are notes about the shift in US fatalities based on the remarkable personal data collection in the late 1800s by Kretzer (1895). A comparison more than a century after Kretzer' summary is made with fatalities during the latest 10 years based on annual information posted annually at www.lightningsafety.noaa.gov. Comparisons in Fig. 8.1 between the two periods show:

- *Agriculture*: The most recent data in Fig. 8.1 indicate that agricultural scenarios of lightning fatalities are now infrequent compared with the 1890s period before the widespread introduction of mechanized farming in the United States.
- *Indoors*: Injuries and deaths indoors was the largest category in the 1890s when people inside homes and other buildings were not safe from the effects of lightning, probably similar to current housing in many lesser-developed countries now. Such situations are no longer the locations of lightning fatalities in the United States. Fatality cases inside dwellings were summarized by Holle (2010). In the United States, there were 21 events resulting in 31 deaths and 4 injuries associated with these events from 1992 to 2010. All but three of the fatalities occurred when a home caught fire at night due to a lightning strike, and deaths occurred to elderly, young, or physically or mentally disabled people.
- *Outdoors*: This has become the largest category during the last decade. Recent routine activities include gardening and mowing in dwellings' yards, walking in neighborhoods, and other everyday outdoor activities vulnerable to lightning. These activities and locations were a smaller fraction of the sample in the 1890s that was dominated by injuries to people inside unsafe structures and engaged in agriculture and other outdoor occupations. Note that the absolute number of people killed while outdoors in the 1890s was larger than it is now, but it was not as large a proportion as at present.
- *Recreation*: This has become the second largest category in the last decade. These scenarios include personal recreational while involved in boating, water

sports, hiking, and camping. In particular, water-related activities have been the largest single activity of all lightning fatalities in the last decade, as indicated at www.lightningsafety.noaa.gov.

- *Small structures*: These are locations such as golf and beach shelters, backyard sheds, and similar very small buildings. Small structures account for a much larger proportion of fatalities in lesser-developed nations than in the United States.
- *Sports*: This category includes organized recreation, such as neighborhood soccer and baseball leagues, golf, as well as sports at all levels of schools.

Questions to explore
Based on this information, how should lightning safety advice and information have changed over the last century in the United States? Is the present emphasis on recreation the correct category?

8.2 Developing Nations

The situation in much of Africa, the Indian subcontinent, Southeast Asia, and other developing regions resembles the lightning vulnerability in developed nations a century ago in the top bar of Fig. 8.1. That is, dwellings and other structures where people currently live and work in developing countries are often not substantial enough to provide protection from lightning's effects. In addition, labor-intensive agriculture remains a major aspect of many people's livelihood and occurs during the daytime when most lightning takes place. For example, 93% of the recent lightning fatalities in Bangladesh are rural (Dewan et al. 2017). Three scenarios are prominent in developing nations.

8.2.1 Agriculture

The profile of activities and locations is similar to the developing world situation many decades ago. A recent study of 445 cases in agricultural situations indicates a large loss of life involving 969 fatalities and 597 nonfatal injuries in Fig. 8.2 (Holle 2016b). Reports from countries with adequate communications indicate that large numbers of people are killed and injured in numerous events during the growing season in India, Bangladesh, and adjacent countries. The large numbers of deaths and injuries per event are very different from the situation in the United States where 90% of all lightning casualties are to one person at a time (Curran et al. 2000). Notable features of these agricultural casualties in mainly India and Bangladesh in Fig. 8.2 include the following:

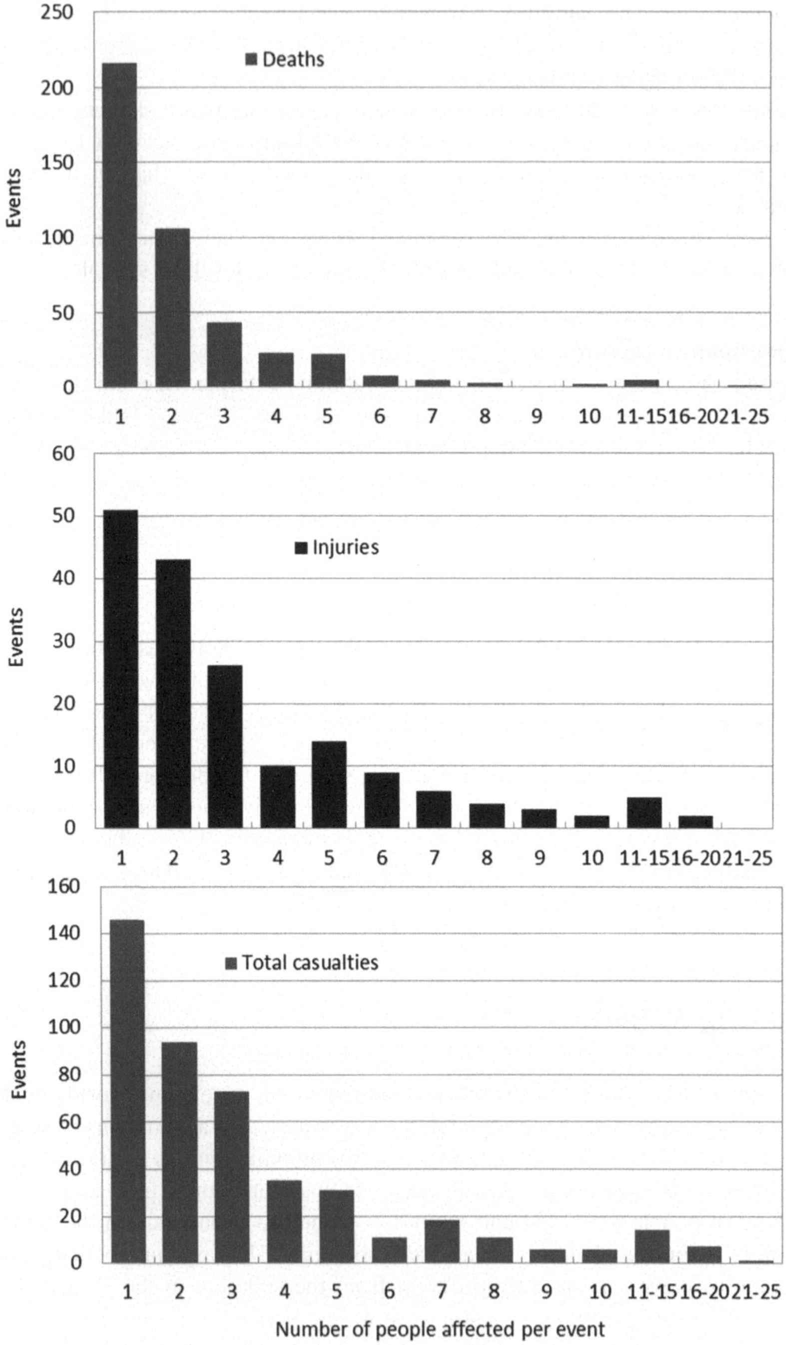

Fig. 8.2 Deaths, injuries, and casualties per agricultural event. (Holle 2016b)

- The proportion of fatalities and injuries is nearly evenly divided between women and men, compared with 83% male in Curran et al. (2000). See Chap. 9 for details.
- Most casualties are during the afternoon.
- Nearly all occur during a storm with rain.

These scenarios indicate that agricultural workers do not stop during daytime labor-intensive field work when lightning is present despite the presence of overhead thunderstorms. In addition, these workers often have no location safe from lightning immediately available. Solutions in agricultural situations include providing buses or other designated lightning-safe structures in the field for groups of people who can quickly reach them when lightning occurs.

8.2.2 Dwellings

Outside the United States, Holle (2010) examined 26 lightning incidents involving dwellings that resulted in 106 deaths and 33 injuries. Another 25 incidents described the dwelling as a hut in the original English-language article emanating from the country of the incident, and those incidents resulted in another 76 deaths and 68 associated injuries.

8.2.3 Schools

A high-profile source of lightning events is the loss of life and multiple injuries occurring in schools as indicated in a summary of 123 cases involving 218 lightning fatalities and 710 nonfatal injuries outside the United States (Holle and Cooper 2016). A portion of these events occurs inside classrooms that are made of mud brick and straw. In developing nations, these buildings have no conducting metal materials in the form of grounded plumbing, wiring, or structural members that take a strike at or near the building and conduct the current safely into the ground without affecting people inside (Fig. 8.3). Another portion of the events at schools occurs while pupils are outside during recess, sporting events, assemblies, lunch, and walking to and from schools. As a result, school-related lightning safety needs to emphasize that students and staff reach a lightning-safe place instead of going under trees or inside lightning-unsafe huts to stay dry. Unfortunately, these are not available in most developing nations. The African Centres for Lightning and Electromagnetics Network (ACLENet) addresses as many of these issues as possible (Cooper et al. 2016). In contrast, there are no cases of people killed by lightning inside schools in the United

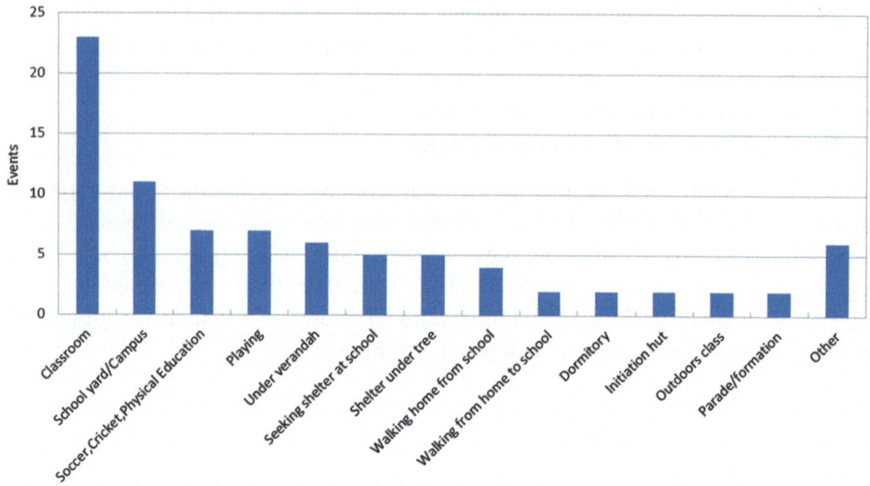

Fig. 8.3 Location and activity of school lightning fatalities and nonfatal injuries (n = 84 reports) from Holle and Cooper (2016)

States, and very few injuries, since they are well-constructed and direct strikes cause no more than a momentary power outage in all but a few cases (Holle 2010).

Question to explore

Based on this information, what is the most efficient way to provide lightning protection to the most people in developing countries?

Answering the call of nature

This situation arises occasionally around the world. In locations with substantial buildings, people have been blown off toilets when a house is struck and a surge of current travels through plumbing. In lesser-developed locations, people may be in small unsubstantial buildings or outdoors in remote areas where serious injury or death can occur. There have been fatalities where people have been in the jungle, fields, or other open spaces.

References

Coates L, Blong R, Siciliano F (1993) Lightning fatalities in Australia, 1824-1993. Nat Hazards 8:217–233

Cooper MA, Gomes C, Tushemereirwe R et al (2016) The development of the African Centres for Lightning and Electromagnetics Network. In: Preprints of the 33rd international conference on lightning protection, Estoril, 25–30 Sept 2016

Curran EB, Holle RL, López RE (2000) Lightning casualties and damages in the United States from 1959 to 1994. J Clim 13:3448–3453
Dewan A, Hossain MF, Rahman MM et al (2017) Recent lightning-related fatalities and injuries in Bangladesh. Weather Clim Soc 9:575–589
Elsom DM (2015) Striking reduction in the annual number of lightning fatalities in the United Kingdom since the 1850s. Weather 70:251–257
Gourbière E (1998) Lightning injury to human beings in France. In: Proceedings of the 24th international conference on lightning protection, Staffordshire University, Birmingham, 14–18 Sept 1998
Holle RL (2010) Lightning-caused casualties in and near dwellings and other buildings. In: Preprints of the 3rd international lightning meteorology conference, Vaisala, Orlando, 21–22 Apr 2010
Holle RL (2016a) A summary of recent national-scale lightning fatality studies. Weather Clim Soc 8:35–42
Holle RL (2016b) Lightning-caused deaths and injuries related to agriculture. In: Preprints of the 33rd international conference on lightning protection, Estoril, 25–30 Sept 2016
Holle RL, Cooper MA (2016) Lightning-caused deaths and injuries at schools. In: Preprints of the 33rd international conference on lightning protection, Estoril, 25–30 Sept 2016
Holle RL, López RE, Navarro BC (2005) Deaths, injuries, and damages from lightning in the United States in the 1890s in comparison with the 1990s. J Appl Meteorol 44:1563–1573
Kretzer HF (1895) Lightning record: a book of reference and information, vol I, St. Louis, p 106
Mills B, Unrau D, Parkinson C et al (2010) Assessment of lightning-related damage and disruption in Canada. Nat Hazards 52:481–499

Chapter 9
Gender, Age, and Casualties per Incident

Abstract The very different fatality rates occurring in a variety of locations and activities in developed versus developing nations were described in Chap. 8. There are often distinct age and gender differences accompanying these incidents. In more developed nations, young males have tended to be the dominant group for many decades, perhaps in part due to great risk-taking behavior and outdoor employment. In lesser-developed nations, limited information suggests a tendency for younger males to make up a disproportionate number of lightning casualties, except in agricultural and school situations that often involve many people per event and where males and females are injured in nearly equal numbers.

9.1 Gender

In more developed nations, the typical scenario in recent years is that over 70% of people killed or injured by lightning are males. Table 9.1 indicates a range from 64% to 87% in various developed nations and the United States. This has been true, to a large extent, for developed countries over the last two centuries. The largest exception is a 49% male ratio of indoor casualties in England and Wales. During the earlier years of the twentieth century, agriculture was considered a major contributor to the higher male occurrence. In recent years, it has also become apparent that a portion of the male dominance is due to their higher level of taking risks. Badoux et al. (2016) emphasize a tendency for more strenuous outdoor recreation, such as alpine climbing, hiking, and camping, that result in male casualties. Categories that are more specific have been compiled in incidents during organized sports and other recreational activities in the United States and other developed nations where these activities take place during leisure recreation. Table 8.1 indicates that a male percentage over 70% also applies for baseball and softball, camping and tenting, golf, and soccer (Holle 2005).

In developing nations of Africa, Southeast Asia, and other regions, only a few national studies include gender. Available data from Bangladesh, Brazil, and Swaziland in the lower portion of Table 8.1 also indicate male ratios of 68% or more.

There are two prominent differentiating issues for developing nations (bottom of Table 8.1). One is the situation of labor-intensive agriculture, where a recent study showed that only 62% of the victims were male (Holle 2016). At schools in Africa

M. A. Cooper, R. L. Holle, *Reducing Lightning Injuries Worldwide*,
Springer Natural Hazards, https://doi.org/10.1007/978-3-319-77563-0_9

Table 9.1 Gender of lightning casualties in developed versus developing nations

	Male percentage	Reference
Developed nations		
Australia	80% of fatalities	Coates et al. (1993)
Canada		
All	72% of fatalities	Mills et al. (2006)
All	77% of injuries	Mills et al. (2006)
England and Wales		
	64% of casualties	Elsom and Webb (2014)
	73% outdoors	Elsom and Webb (2014)
	49% indoors	Elsom and Webb (2014)
	83% of fatalities	Elsom and Webb (2014)
Greece	86% of fatalities	Agoris et al. (2002)
Poland	75% of fatalities	Loboda (2008)
Singapore	82% of fatalities	Pakiam et al. (1981)
United States		
All	84% of fatalities	Curran et al. (2000)
All	82% of injuries	Curran et al. (2000)
Colorado	76% of casualties	López et al. (1994)
Florida	87% of fatalities	Duclos et al. (1990)
Florida	61% of fatalities	Duclos et al. (1990)
Florida	87% of casualties	Holle et al. (1993)
Primarily United States		
Baseball and softball	71% of casualties	Holle (2005)
Camping and tenting	76% of casualties	Holle (2005)
Golf	94% of casualties	Holle (2005)
Soccer	91% of casualties	Holle (2005)
Developing nations		
Bangladesh	80% of fatalities	Dewan et al. (2017)
Brazil	81% of fatalities	Cardoso et al. (2014)
Swaziland	68% of fatalities	Dlamini (2008)
Primarily India and Bangladesh		
Agriculture	62% of casualties	Holle (2016)
Primarily Africa and India		
Schools	50% of casualties	Holle and Cooper (2016)

and India, females account for 50% of the deaths and injuries (Holle and Cooper 2016). At schools, and for at least lower grades, the students tend to be evenly divided between the sexes so that casualties would be expected to be nearly equal as well.

Questions to explore
Identify situations where male risk-taking is evident in addition to being less responsive to the risk of lightning. How can such behavior be changed with regard to lightning?

9.2 Age

Age distributions of lightning casualties have not been reported as often as the gender. In addition, age categories differ among studies so that direct comparisons such as percent male in the previous section cannot be made readily. Nevertheless, Table 9.2 indicates a trend toward younger ages that is partially indicative of the relatively young population in many developing countries.

In developed nations, those between 18 and 30 account for about half of all casualties in Australia, Canada, France, Singapore, and the United States (Table 9.2). There are exceptions to this younger male generality, most often in recent years in hiking and climbing incidents. These activities have a wider age range between 11 and 60 than the previous examples (Holle 2005). The compilation in Table 9.2 includes events both in the United States and in other developed nations where these activities take place during leisure recreation.

Table 9.2 Ages of lightning casualties in developed versus developing nations

	Age distribution	Reference
Developed nations		
Australia	Peak from 15 to19 years old	Coates et al. (1993)
Canada		
All	51% of fatalities from 16 to 45	Mills et al. (2006)
All	61% of injuries from16 to 45	Mills et al. (2006)
France	61% of fatalities from 15 to 44	Gourbière (1999)
Singapore	58% of fatalities from 10 to 29	Pakiam et al. (1981)
Switzerland	23% of fatalities from 10 to 19	Badoux et al. (2016)
United States		
All	Highest death rate from 20 to 34	Thacker et al. (2008)
Colorado	49% from 16 to 35	López et al. (1994)
Florida	45% of casualties from16 to 35	Holle et al. (1993)
Primarily United States		
Baseball and softball	69% of casualties from 11 to 20	Holle (2005)
Camping and tenting	53% of casualties from 11 to 20	Holle (2005)
Golf	62% of casualties from 31 to 55	Holle (2005)
Soccer	62% of casualties from 11 to 15	Holle (2005)
Developing nations		
Bangladesh	69% of fatalities from 10 to 39	Dewan et al. (2017)
Brazil	43% of fatalities from 20 to 39	Cardoso et al. (2014)
Swaziland	67% of fatalities from 10 to 39	Dlamini (2008)
Primarily India and Bangladesh		
Agriculture	51% of casualties from 16 to 35	Holle (2016)
Primarily Africa and India		
Schools	66% of casualties from 12 to 16	Holle and Cooper (2016)

In developing nations, only a few national studies include age (Table 9.2). For Bangladesh, Brazil, and Swaziland, there is a dominant grouping between the ages of 10 and 39. There are two specific issues to address for developing nations. One is the situation of labor-intensive agriculture, where a recent study showed that half of the casualties were between 16 and 35, while the median age of female casualties tended to be about a decade older than males (Holle 2016). Secondly, in school incidents, due to the typical school victims being in primary and secondary schools, the corresponding ages are in the teens (Holle and Cooper 2016).

Questions to explore

Why is the age range of the mid-teens to middle 30s so common in lightning deaths and injuries? Is this age range likely to change in developing nations as their economic situation changes?

All of the above summaries report the actual number of casualties by age. However, they are not weighted by the population within each age group except where this ratio was calculated for Bangladesh by Dewan et al. (2017). Figure 9.1a shows the number of people killed by lightning in each age range. Figure 9.1b indicates that although weighting by the population in each group has the same general features as 9.1a, there is a shift in rates. For males, population weighting shifts the maximum to the 30–39 age range and indicates a larger relative frequency at ages over 60 than the number of fatalities. For females, there is also an indication of more deaths per population over age 50 than indicated by fatalities in Fig. 9.1a. That is, people over 50 are lightning casualties proportionally more often than is indicated by the actual numbers of casualties in other age groups. Two-thirds of the Bangladesh incidents included in this summary involve farming and being inside houses (Dewan et al. 2017).

9.3 Casualties per Incident

An indicator related to gender and age is represented by the grouping of people in incidents. There is a strong distinction in the number of people killed and injured by lightning per event between developed and developing nations. Table 9.3 shows that in the United States from 1959 to 1994, 91% of lightning casualty events had only one person killed, and 68% had a single person injured (Curran et al. 2000). Limited data are available about casualties per incident in other developed nations; England and Wales, Singapore, and the state of Colorado in the United States also have over 66% single-person incidents.

However, multiple rather than single casualties occur more commonly in the recreational activities noted in Table 9.3 (Holle 2005). The rate is as high as 69% multiple deaths and injuries for soccer; quite a few of these cases occurred outside

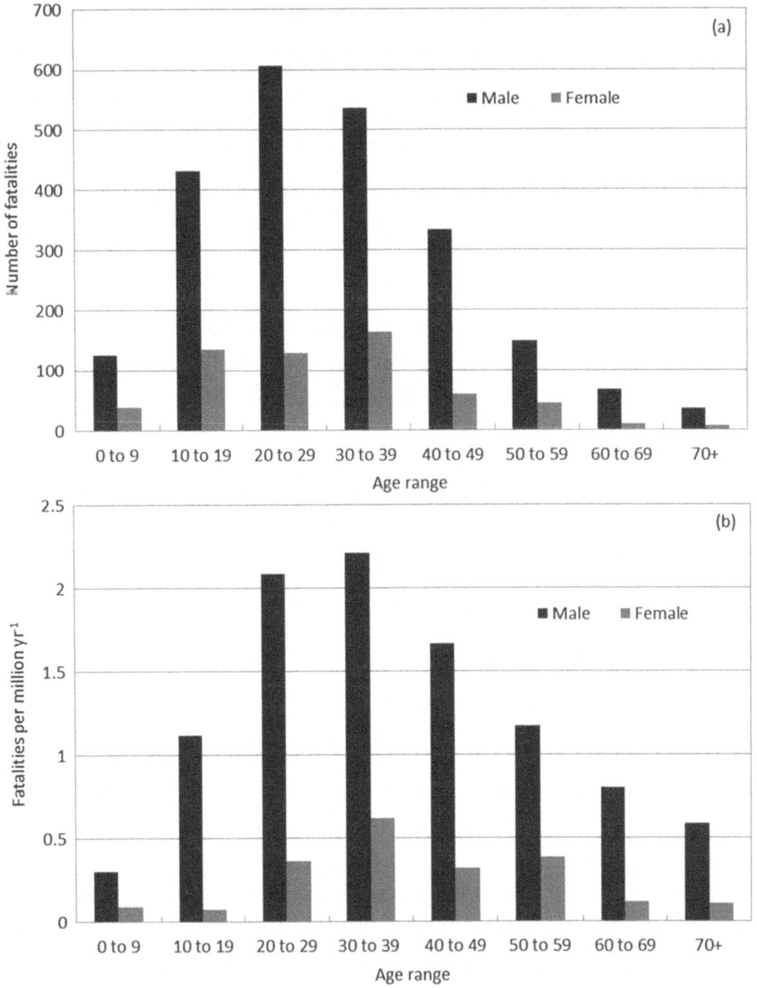

Fig. 9.1 Age distribution by gender of (**a**) lightning-related fatalities and (**b**) population-weighted fatalities per year. (Dewan et al. 2017)

the United States. The tendency for leisure sports to have more than one casualty may also result in more frequent reporting due to their situation seemingly being an activity related to pleasure rather than employment.

Developing nations have a tendency for multiple casualties per incident in agricultural and school events. Note in Table 9.3 that injuries in schools are almost all multiple-casualty incidents (92%). Lightning incidents at schools can involve up to 20 or more casualties at a time and 50 or more people in agricultural events. There are frequent agricultural events (Holle 2016) that involve nearly all females with many deaths and injuries in the situation of crews working in a field, often in the age range of the 30s or more.

Table 9.3 Number of lightning casualties per incident in developed versus developing nations

	Casualties per incident	Reference
Developed nations		
England and Wales	98% single fatalities	Elsom and Webb (2014)
Singapore	72% single fatalities	Pakiam et al. (1993)
United States		
All	91% single fatalities	Curran et al. (2000)
All	68% single injuries	Curran et al. (2000)
Colorado	66% single casualties	López et al. (1995)
Primarily United States		
Baseball and softball	43% single casualties	Holle (2005)
Camping and tenting	36% single casualties	Holle (2005)
Golf	37% single casualties	Holle (2005)
Soccer	31% single casualties	Holle (2005)
Developing nations		
Bangladesh	50% single fatalities	Dewan et al. (2017)
Swaziland	78% single fatalities	Dlamini (2008)
Primarily India and Bangladesh		
Agriculture	50% single fatalities	Holle (2016)
Agriculture	29% single injuries	Holle (2016)
Primarily Africa and India		
Schools	39% single fatalities	Holle and Cooper (2016)
Schools	8% single injuries	Holle and Cooper (2016)

Questions to explore

Does the ready availability of substantial buildings and fully enclosed metal-topped vehicles in developed regions of the world lead to more single-casualty events than in developing countries? Since multiple-fatality cases are reported more often in developing regions, are these data reliable?

9.4 Summary

To provide perspective with regard to extreme events, several incidents with multiple casualties since 1873 can be listed as follows:

- *Agriculture*: The largest known number of agricultural fatalities in a single incident was 22 farmers who died while planting seeds near Nanchang, China, in May 1994 (Grazulis 1996).

- *Dwelling*: The largest known loss of life from a single dwelling event was 21 fatalities when people sought shelter from rain in a hut in the Manica Tribal Trust Lands of present-day Zimbabwe in 1975 (Cerveny et al. 2017).
- *Indirect*: The largest known loss of life from an indirect strike was 469 fatalities due to a lightning-caused oil tank fire in Dronka, Egypt, in 1994 (Cerveny et al. 2017).
- *School*: The largest known single school case involved 7 deaths and 67 nonfatal injuries in 2010 at a South African kindergarten (Holle and Cooper 2016). The case with the largest number of deaths was at a school in Uganda in 2011 where 18 children were killed and 38 were admitted to the hospital. See http://www.telegraph.co.uk/news/weather/8606238/Lightning-strike-kills-18-children-in-Uganda.html.

References

Agoris D, Pyrgioto E, Vasileiou D et al (2002) Analysis of lightning death statistics in Greece. In: Proceedings of the 26th international conference on lightning protection, Cracow, 2–6 Sept 2002

Badoux A, Andres N, Techel F et al (2016) Natural hazard fatalities in Switzerland from 1946 to 2015. Nat Hazards Earth Syst Sci 16:2747–2768

Cardoso I, Pinto O Jr, Pinto IRCA et al (2014) Lightning casualty demographics in Brazil and their implications for safety rules. Atmos Res 135–136:374–379

Cerveny R, Bessemoulin P, Burt CC et al (2017) WMO assessment of weather and climate mortality extremes: lightning, tropical cyclones, tornadoes, and hail. Weather Clim Soc 9:487–497

Coates L, Blong R, Siciliano F (1993) Lightning fatalities in Australia, 1824-1993. Nat Hazards 8:217–233

Curran EB, Holle RL, López RE (2000) Lightning casualties and damages in the United States from 1959 to 1994. J Clim 13:3448–3453

Dewan A, Hossain MF, Rahman MM et al (2017) Recent lightning-related fatalities and injuries in Bangladesh. Weather Clim Soc 9:575–589

Dlamini WM (2008) Lightning fatalities in Swaziland. Nat Hazards. https://doi.org/10.1007/s11069-008-9331-9336

Duclos PJ, Sanderson LM, Klontz KC (1990) Lightning-related mortality and morbidity in Florida. Public Health Rep 105:276–282

Elsom DM, Webb JDC (2014) Deaths and injuries caused by lightning in the UK, 1988-2012. Weather 69:221–226

Gourbière E (1999) Lightning injuries to humans in France. In: Proceedings of the 11th international conference on atmospheric electricity, NASA/CPP-1999-209261, Guntersville, 7–11 June 1999

Grazulis TP (1996) Significant tornadoes update, 1992-1995. Environmental Films, St. Johnsbury

Holle RL (2005) Lightning-caused recreation deaths and injuries. In: Preprints, of the 14th symposium on education, American Meteorological Society, San Diego, 9–13 Jan 2005

Holle RL (2016) Lightning-caused deaths and injuries related to agriculture. In: Preprints of the 33rd international conference on lightning protection, Estoril, 25–30 Sept 2016

Holle RL, Cooper MA (2016) Lightning-caused deaths and injuries at schools. In: Preprints of the 33rd international conference on lightning protection, Estoril, 25–30 Sept 2016

Holle RL López RE, Ortiz R et al (1993) The local meteorological environment of lightning casualties in Central Florida. In: Preprints of the 17th conference on severe local storms and conference on atmospheric electricity, American Meteorological Society, St. Louis, 4–8 Oct1993
Loboda M (2008) Lightning deaths and injuries in Poland in period of 2001-2006. In: Preprints of the 29th international conference on lightning protection, Uppsala, 23-26 June 2008
López RE, Holle RL, Heitkamp TA (1994) Deaths, injuries, and property damage due to lightning in Colorado from 1950 to 1991 based on Storm Data. NOAA Tech Memo NSSL-103, Natl Severe Storms Lab
López RE, Holle RL, Heitkamp TA (1995) Lightning casualties and property damage in Colorado from 1950 to 1991 based on storm data. Wea Forecasting 10:114–126
Mills B, Unrau D, Parkinson C et al (2006) Striking back: an assessment of lightning-related fatality and injury risk in Canada. Final Tech Report, Environment Canada
Pakiam J, Chao T, Chia J (1981) Lightning fatalities in Singapore. Meteor Mag 110:175–187
Thacker MF, Lee R, Sabogal RI et al (2008) Overview of deaths associated with natural events, United States, 1979–2004. Disasters 32:303–325

Chapter 10
Contributors to Lightning Casualty Risk

Abstract Myths involving all types of weather phenomena are widespread and often based on personal experiences of individuals and societies through time. Lightning is especially prone to stories that attribute survival to what a person was wearing, holding, or how they were positioned at the time of the strike. This perspective is misapplied, since about 90% of people affected by lightning survive in developed regions of the world.

The odds of lightning affecting an individual are falsely perceived to be very low, frequently resulting in complacency with regard to the threat. In addition, recreational activities may continue during thunderstorms that have low flash rates, making it more difficult to appreciate the danger. Subsistence agricultural activities in less developed regions, on the other hand, tend to continue during heavy rain regardless of its attendant lightning.

10.1 Myths

All types of weather phenomena have myths attached to them. Storms perceived as unusual may be said not to have occurred in the "living memory" of anyone in a village or neighborhood, although long-term records will show that it has a similar precedent. In tropical regions around the world, cyclones are said not to occur in certain places because of locally made-up stories based on interpretations with limited experience since they occur infrequently. Tornadoes in the United States are sometimes said to miss cities or not cross rivers, which are entirely false statements. Past snowstorms are remembered as stronger than ever, potentially because of a strong impact made by a singular event in childhood that is enhanced by intervening years. These opinions are based on personal or local community short-term memories of past events that become a type of rule with the passage of time, despite documented weather evidence to the contrary. Such perceptions are developed because some events are infrequent with long intervals between impacts at the immediate location. However, when objective data from such phenomena as floods or hurricanes are examined, there is almost always a precedent for a similar event.

M. A. Cooper, R. L. Holle, *Reducing Lightning Injuries Worldwide*,
Springer Natural Hazards, https://doi.org/10.1007/978-3-319-77563-0_10

One of the pernicious myths is the mistaken perception that the weather at a location is entirely due to what is visible from where a person stands. That approach results in the thought process that a small nearby hill causes storms to form, die, or miss the location or a small pond provides the moisture for a thunderstorm that crosses a local region (Chap. 13). Instead, meteorological factors on the global, national, and regional scales influence what occurs locally. Wind velocities and directions change constantly at all levels of the atmosphere through time in a never-ending sequence that is not necessarily apparent when viewed from a single location.

The list of such myths and misconceptions regarding meteorological events is nearly endless. These stories are common in developed countries, in less developed nations, in cities, and in agricultural areas. In fact, every person in the world can be said to have a set of preconceived notions based on their life history of weather experiences.

Lightning, in particular, has an extremely large range of myths. Survivors' stories take on a sense of credible fact when all of the irrelevant circumstances of a specific event are associated with survival. In the developed countries of the world, about 90% of those affected by lightning survive, and some are not greatly impacted (Cherington et al. 1999). Thus, what a person happened to be wearing, holding, or their location or posture is generalized from the singular lightning event (Roeder 2007; Trengove and Jandrell 2010). The list of irrelevant aspects of that moment can be very large. For example, in parts of Africa, it is said that wearing red attracts lightning, which is based on some long-forgotten, often-repeated, and mistakenly understood incident. Unfortunately, these features are overextended with the passage of time to become factual lightning safety advice that is told to others. In reality, there is no relationship between those factors making one safe from lightning, and they cannot be taken as any evidence of a specific posture, location, or other feature as always being applicable. Extending a single scenario to a generality can result in critical errors in safety advice. For this reason and to be scientifically supported, conclusions should be based on summarizing as large a sample of cases over as long a period and as large an area as possible,

Questions to explore
What weather beliefs are common in your area? How would you be able to determine which are false? What pitfalls may occur if you try to tell others that a strange or long-held impression about lightning safety is incorrect? How do you tell family members the correct information?

10.2 Odds

Nearly everyone in the world hears thunder emanating from lightning at least once every year. Yet not many people are killed or injured, so the low frequency of direct personal knowledge of lightning fatalities and injuries leads to a conclusion that the

Table 10.1 Odds of being a lightning casualty in the United States

Assumption	Resulting odds
Population of 330,000,000 people	
30 killed per year (*Storm Data*)	1 in 11,000,000
300 injured per year *(Storm Data)*	1 in 1,100,000
330 killed and injured per year *(Storm Data)*	1 in 1,000,000
Life span of 80 years	1 in 12,500
Major impact on 10 people	1 in 1250

lightning threat can be dismissed. The true threat is actually more extensive than is perceived. Table 10.1 considers the example of the United States. Numbers have been adjusted slightly to make the odds result in round numbers. In summary, the odds are about one in 1250 over an 80-year life span that a person in the United States will be killed or injured by lightning or be a close relative or friend of a victim at today's casualty rate. These very broad generalizations are for instructional purposes only, for a large country, and do not take into account regional variations.

In developing countries, the odds are less favorable than shown for the United States in Table 10.1. For example, over 50% of Lake Victoria (Africa) community members who were interviewed were aware of at least one person who had been injured due to lightning on the lake in the past year (Tushemereirwe et al. 2017). The survival rate of those affected by lightning is likely lower than the 90% reported by Cherington et al. (1999) for the United States. Many of the casualties in developed countries involve less severe impacts due to people being inside lightning-safe buildings and vehicles. Instead, numerous multiple casualty incidents may occur in developing countries because of the lack of substantial buildings and nearby lightning-safe vehicles (Sect. 6.5). The lack of rescue systems and good medical care further increases the chances of death.

Another contributor to lightning casualties is the perception that the odds are somehow suppressed during certain activities. In developed nations, outdoor recreation during weekends or holidays may be planned long in advance, so that when the mountain hike or beach outing finally occurs, there is a rationalization that lightning occurrence during leisure time is somehow less likely and can be ignored (Hodanish et al. 2004). Similarly, in lesser-developed regions, the need to continue daytime labor-intensive subsistence agriculture may result in a continuation of outdoor work despite the presence of lightning (Holle 2016).

Questions to explore

How do you perceive your lightning threat? What list of factors could be entered into a list similar to those in Table 10.1 to identify the odds in your country?

10.3 Rain

In humid locations, people usually attempt to find a dry place during rainfall with trees (Holle 2012) and small unsubstantial structures (Holle 2010) often being the most convenient locations. However, staying dry is not the same as being safe from lightning. Staying safe from lightning injury is further complicated because:

1. Lightning often occurs before the storm and before the rain begins (Holle et al. 1993; Lengyel et al. 2005),
2. Lightning may occur up to 15 km from the edge of the heavy rain area at the ground (Rison et al. 2003),
3. Some will leave a safe area too soon after the rain ends and be exposed to lightning strikes (Holle et al. 1993; Lengyel et al. 2005).

In dry environments such as deserts and high mountains, there may be little or no rainfall in the presence of lightning. Many cases of lightning occurring overhead when virtually no rainfall is occurring have been documented where it is often called a "bolt from the blue" (Lengyel et al. 2005; Hodanish et al. 2015). These are especially difficult situations to manage due to the frequent occurrence of a very short time lag from the first lightning to those causing fatalities or injuries.

Finally, lightning death and injury events can be divided equally among small, moderate, and large rates of lightning at the time of injury (Holle et al. 1993). Recreational events in the United States may take place during times when the lightning rate is small and is accompanied by light rain or no rain at all, at the time and location of the death or injury (Hodanish and Zajac 2002; Hodanish et al. 2004, 2015). These incidents with small flash rates often reinforce the perception that the storm is not dangerous since it may not be raining heavily enough to seek a dry place. In contrast, agricultural events in developing nations often occur in heavy rain due to the time and financial pressure of planting and harvesting requirements with subsistence farming (Holle 2016).

Questions to explore

Under what situations do workers continue to stay outside when it's raining and lightning is imminent? How do you manage the threat of very low lightning rate storms?

References

Cherington M, Walker J, Boyson M et al (1999) Closing the gap on the actual numbers of lightning casualties and deaths. In: Preprints of the 11th conference on applied climatology, American Meteorological Society, Dallas, 10–15 Jan 1999

Hodanish S, Zajac B (2002) Documentation of the "first lightning flash of the day" associated with a weak shallow convective updraft killing an 18 year old on top of pikes peak, Colorado.

In: Preprints of the 17th international lightning detection conference, Vaisala, Tucson, 16–18 Oct 2002

Hodanish S, Holle RL, Lindsey DT (2004) A small updraft producing a fatal lightning flash. Weather Forecast 19:627–632

Hodanish S, Wolyn P, Mozley K (2015) Meteorological analysis of the Rocky Mountain National Park lightning fatalities of 11 and 12 July, 2014. In: Preprints of the 7th conference on the meteorological applications of lightning data, American Meteorological Society, Phoenix, 4–8 Jan 2015

Holle RL (2010) Lightning-caused casualties in and near dwellings and other buildings. In: Preprints of the 3rd international lightning meteorology conference, Vaisala, Orlando, 21–22 Apr 2010

Holle RL (2012) Lightning-caused deaths and injuries in the vicinity of trees. In : Preprints of the 31st international conference on lightning protection, Vienna, 2–7 Sept 2012

Holle RL (2016) Lightning-caused deaths and injuries related to agriculture. In: Preprints of the 33rd international conference on lightning protection, Estoril, 25–30 Sept 2016

Holle RL, López RE, Ortiz R et al (1993) The local meteorological environment of lightning casualties in Central Florida. In: Preprints of the 17th conference on severe local storms and conference on atmospheric electricity, American Meteorological Society, St. Louis, 4–8 Oct 1993

Lengyel MM, Brooks HE, Holle RL et al (2005) Lightning casualties and their proximity to surrounding cloud-to-ground lightning. In: preprints of the 14th symposium on education, American Meteorological Society, San Diego, 9–13 Jan 2005

Rison W, Krehbiel P, Thomas R et al (2003) Lightning mapping and radar observations of bolts from the blue. In: Proceedings of the 12th international conference on atmospheric electricity, Versailles, 9–13 June 2003

Roeder WP (2007) Pernicious lightning myths. In: Preprints of the international conference on lightning and static electricity, Paris, 28–31 Aug 2007

Trengove E, Jandrell IR (2010) Strategies for understanding lightning myths and beliefs. In: Preprints of the 30th international conference on lightning protection, Cagliari, 13–17 Sept 2010

Tushemereirwe R, Cooper MA, Tuhebwe D et al (2017) The most effective methods for delivering severe weather early warnings to fishermen on Lake Victoria. PLOS Curr Disasters, February 22. https://doi.org/10.1371/currents.dis.d645f658cf20bc4a23499be913f1cbe1

Part III
When and Where Lightning Occurs

Chapter 11
Global Lightning Distribution

Abstract Cloud-to-ground lightning occurs almost everywhere around the globe, although it comes to ground more often at some times and locations than others. Lightning is ubiquitous, unlike tornadoes that are relatively rare and cover small areas and tropical cyclones that are confined to coastal areas and some distance inland. This chapter will provide information on when and where lightning occurs and at what frequency, an important aspect of reducing lightning fatalities worldwide. The underlying meteorological factors that result in the lightning phenomenon are the same globally (Chap. 13). In contrast, people's vulnerability and their reaction to the occurrence of lightning vary, depending on the social and economic situation of each country (Chap. 6).

11.1 US Cloud-to-Ground Lightning

There has been extensive experience with measuring the occurrence of cloud-to-ground lightning in the United States for over 30 years. The National Lightning Detection Network (NLDN) has been collecting data with ever-improving capability during this period (Cummins and Murphy 2009; Nag et al. 2015). The US NLDN maps have been published for over two decades starting with Orville (1991) through the most recent maps by Orville et al. (2011) and Holle et al. (2016).

The 10-year NLDN map in Fig. 11.1 illustrates how cloud-to-ground flash density varies by more than two orders of magnitude across the continental United States. The largest cloud-to-ground flash density is in Florida and along the coast of the Gulf of Mexico, exceeding 12 flashes/square km/year in some areas. In this area, warm ocean water is adjacent to land that is strongly heated during warm-season afternoons, resulting in both heavy rain and frequent lightning. In contrast, there are locations along the west coast of the United States where flashes are so infrequent that an average of only one flash/year has been detected in some 20 by 20 km grid squares. This minimal flash occurrence is due to cool offshore water and large-scale subsidence aloft that is typical of the west coasts of continents at this latitude.

Over the United States, lightning generally decreases north and westward from the Gulf of Mexico and western Atlantic-Caribbean basin that provides much of the

M. A. Cooper, R. L. Holle, *Reducing Lightning Injuries Worldwide*,
Springer Natural Hazards, https://doi.org/10.1007/978-3-319-77563-0_11

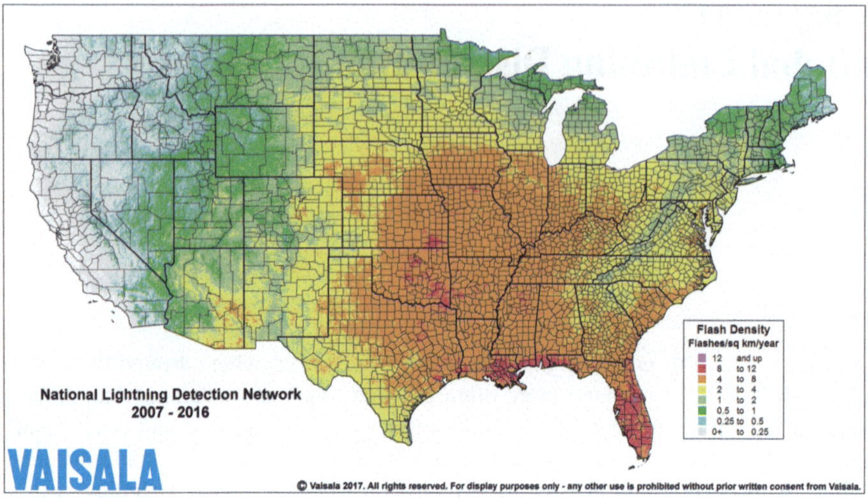

Fig. 11.1 Annual cloud-to-ground flash density in flashes per square km per year over the continental United States based on an average of 22 million flashes per year from the National Lightning Detection Network from 2007 through 2016. Scale in lower right

deep layer moisture needed for thunderstorm formation (Chap. 13). The amount of atmospheric liquid water content in the vertical decreases away from these sources, due in part to the decrease in the number of days with deep moisture arriving at locations further north and west from the Gulf-Atlantic-Caribbean source region. Major features of lightning occurrence over the United States include the following factors that also exist in other areas of the world:

- Large flash densities along the Florida and Gulf of Mexico coasts are due primarily to daytime sea breezes.
- The center of the United States has large flash densities due to a combination of traveling cold fronts, squall lines, upper- and low-level troughs, mesoscale convective systems, and subtropical and tropical systems during the course of the year.
- The Appalachian Mountains in the eastern portion of the map have a northeast-southwest minimum shown in green due to thunderstorms beginning at higher elevations. In this situation, the lower-level moisture-rich atmosphere that is a main contributor to strong upward motion is not as deep as on either side of the elevated terrain.
- In the western third of the United States, surface topography dominates the locations of lightning due to diurnally forced cycles of thunderstorm formation. There are large-scale gradients in terrain elevation east of the Rocky Mountains, medium-sized escarpments in central Arizona and New Mexico, and small-scale features near more isolated mountain ranges in interior states.

Networks similar to the NLDN have been established in over 40 countries (Chap. 14). However, they often are not connected with each other. For that reason, maps are usually not available for more than one country at a time over a period of many years from national networks such as the NLDN.

What is a typical cloud-to-ground flash density for the 48 contiguous United States? Is the east-west variation larger than the north-south changes?

11.2 Global Lightning Overview

The global distribution of lightning is depicted with data from the ground-based Global Lightning Dataset GLD360 in Fig. 11.2 and a combination of two satellite-based sensors for differing periods in Fig. 11.3. GLD360 primarily detects cloud-to-ground lightning while the satellite sensors tend to detect more in-cloud lightning than GLD360. Chapter 14 discusses lightning detection in more detail.

It is apparent from both ground- and satellite-based measurement techniques that lightning is concentrated over land compared with water and is most frequent in equatorial regions compared with higher latitudes. In general, there is more lightning in the Northern Hemisphere than the Southern because of the larger landmass to the north of the equator.

Largest concentrations of lightning are along ocean coasts near warm surface water, which occurs on the east coasts of continents at lower latitudes such as Florida. This is also true for eastern China, where warm ocean water flows northward from the tropics and provides substantial amounts of atmospheric water vapor

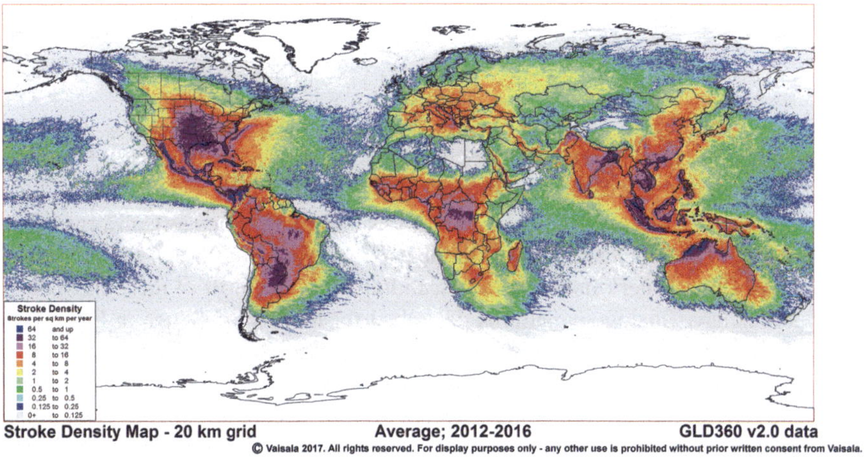

Fig. 11.2 Annual lightning density in strokes per square kilometer per year over the globe based on a total of 7,828,464,140 strokes from the Global Lightning Dataset GLD360 network from 2012 through 2016. Scale in lower left of map

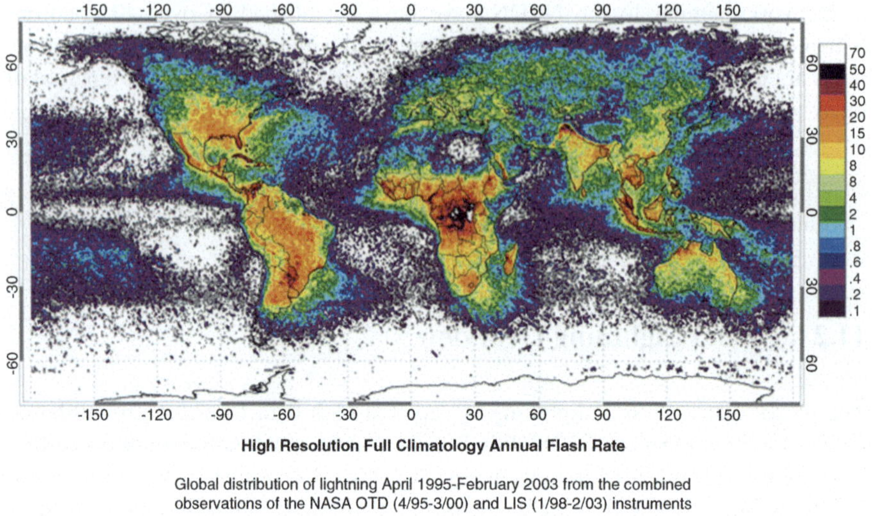

Fig. 11.3 Annual flash rate from combined observations based on data from the NASA Optical Transient Detector (April 1995 through March 2000) and Lightning Imaging Sensor (January 1998 through February 2003). Scale on right side of map

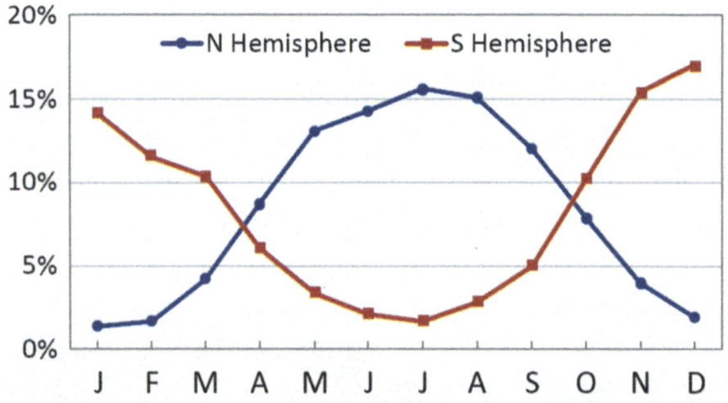

Fig. 11.4 Lightning stroke percentages by month for the Northern and Southern Hemispheres based on GLD360 data in Fig. 11.2. (Holle et al. 2017)

above the surface. In contrast, west coasts at lower latitudes, such as California, and the west coasts of Africa and South America are locations of minimal lightning due to cool upwelling water offshore and descending air aloft. Most of these flash density features have analogous patterns described for the United States in Fig. 11.1. Large lightning densities are also located along the slopes of major elevation changes such as the Andes and Himalaya mountains and equatorial Asian islands. Very small lightning frequencies are detected over the oceans away from land, especially at higher latitudes. The land areas of Antarctica and Greenland have no lightning, and the Sahara has minimal activity.

The difference between lightning in the two hemispheres is apparent in Fig. 11.4 that shows the percent of all lightning detected by month (Holle et al. 2017). In the Northern Hemisphere, activity peaks in the months of May through September, while the Southern Hemisphere has the most lightning in its summer months of November through March (details in Chap. 12).

11.3 Africa

The annual movement of the equatorial trough, also called the intertropical convergence zone, has a major impact on lightning over the continent. It moves northward to the southern edge of the Sahara Desert during the Northern Hemisphere summer and moves southward into the Southern Hemisphere when it is summer there. Some specific features over Africa in Figs. 11.2 and 11.3 are the following:

- One of the world's lightning maxima is in the eastern Democratic Republic of the Congo (Albrecht et al. 2016). Here, the equatorial trough persists over the same location in its annual cycle more often than elsewhere. There is also an important large topographic gradient in this region to the west of Lake Victoria (Holle and Murphy 2017).
- To the north of the region, where the equatorial trough migrates to its northernmost latitude, the Sahara has minimal lightning all year.
- From the western tip of Africa, from Senegal south and east to Nigeria, traveling tropical (easterly) waves move from east to west during the Northern Hemisphere summer as indicated by lightning extending offshore to the west over the North Atlantic Ocean. These tropical waves are the source of some of the tropical storms and hurricanes that subsequently form and travel across the Atlantic Ocean to reach North America.
- Over the southern portion of Africa, disturbances are in the middle latitudes and generally travel from west to east, as indicated by lightning occurrence extending offshore to the east of South Africa over the South Indian Ocean.
- Note the coastal maximum in the Middle East along the western shore of Saudi Arabia and Yemen, where intense land heating and the adjacent Red Sea provides low-level moisture to cause afternoon thunderstorms to form.

Questions to explore
What is the lightning density at your location? How does the lightning density map correspond with the lightning fatality rates in countries where it has been reported (Fig. 5.1)?

11.4 Asia

Mountainous island chains surrounded by very warm ocean waters are major factors resulting in many areas of large lightning densities over Southeast Asia including Malaysia, Indonesia, the Philippines, and adjacent nations. They have lightning dominated by diurnal effects during much of the year (Holle and Murphy 2017). Specific features over Asia that can be noted in Figs. 11.2 and 11.3 are:

- The Indian Monsoon has a significant impact on the timing of lightning over much of Asia (Nag et al. 2017). Thunderstorms and heavy rain begin to the south in May and sweep northward to cover all of India and surrounding countries by the middle of July, then retreat southward again in the autumn months of the Northern Hemisphere.
- On the Indian subcontinent, the Himalayas provide a sharp boundary to the north. Moisture sources at the lower levels that are needed to fuel thunderstorms at the high elevations north of the Himalayas are limited, resulting in minimal lightning occurring over large regions of western China and adjacent countries.

11.5 Australia

The northern region of Australia, closest to the equator, has the same factor of heated land adjacent to warm waters as noted for Florida and the Gulf Coast in the United States and the large island nations in Asia. Lightning density decreases inland away from the tropical oceanic moisture source (Figs. 11.2 and 11.3). The southern half of the country is in the middle-latitude range of westerly winds as indicated by an enhanced region of lightning extending offshore to the east of Australia over the South Pacific Ocean.

11.6 Europe

Due to the higher latitude of Europe, lightning generally occurs less often than on other continents (Figs. 11.2 and 11.3). Most of Europe is in the middle-latitude westerlies, and lightning is due to the traveling cold fronts, squall lines, upper- and lower-level troughs, and mesoscale convective systems as in the Central United States. Lightning mostly occurs during the summer and is less diurnally dependent than in many other regions of the globe. The largest lightning densities are around the Mediterranean shoreline, over and adjacent to Italy, where there are large elevation changes near the water. However, the Mediterranean Sea is not nearly as warm as seas in more tropical regions, so thunderstorms are infrequent over the water.

11.7 North America

The US portion of Figs. 11.2 and 11.3 is discussed in Sect. 11.1. Some additional features to note are:

- Lightning decreases to the north over Canada. Nevertheless, some lightning has been detected as far as 75 degrees north during a few summer days.
- The Rocky Mountains in western Canada provide a boundary between Alberta and British Columbia that blocks low-level, northward flowing atmospheric moisture from reaching the west coast.
- Cuba and Hispaniola have strong lightning maxima due to large mountains surrounded by warm tropical waters, in a manner similar to Southeast Asia.
- Mexico and Central America are locations of strong coastal lightning activity. The North American lightning maximum is in northwest Mexico on the slope of the Sierra Madre Occidental mountain range adjacent to very warm offshore waters (Holle and Murphy 2015).
- Many of the small islands of the eastern and southern Caribbean have almost no lightning due to strong easterly flow throughout the year, shallow atmospheric moisture that minimizes the development of deep afternoon convection, and relatively low elevations compared with the mountainous Asian island chains.

11.8 South America

The northwest tip of South America has lightning frequency rivaling that of East-Central Africa (Albrecht et al. 2016). Specific features in Figs. 11.2 and 11.3 are:

- There is a nighttime maximum over Lake Maracaibo due to it being surrounded by high mountains (Holle and Murphy 2017).
- Northern Brazil is dominated nearly year-round by afternoon tropical convection that moves slowly from east to west.
- The Andes provides a tall boundary that prevents convection over the coastal regions of Chile and Peru and the South Pacific Ocean since deep moisture cannot flow westward over the mountains.
- Over southern Brazil, northern Argentina, Uruguay, and Paraguay, lightning occurs predominantly in the Southern Hemisphere summer. The same factors apply as in the Central United States and Europe, those of traveling cold fronts, squall lines, upper- and lower-level troughs, and mesoscale convective systems. The latter are especially prevalent over northern Argentina and produce very frequent lightning, most often at night. At these higher latitudes, lightning extends offshore from west to east over the Atlantic Ocean.

Questions to explore

Which continent has the most lightning? Why? Which continent has the largest lightning density (most lightning per area)? Why? What regions and countries of the world have frequent lightning that overlaps high population density? Where do traveling weather systems have a greater impact on lightning than other locations?

References

Albrecht R, Goodman SJ, Buechler DE et al (2016) Where are the lightning hotspots on Earth? Bull Am Meteorol Soc 97:2051–2068

Cummins KL, Murphy MJ (2009) An overview of lightning locating systems: history, techniques, and data uses, with an in-depth look at the U.S. NLDN. IEEE Trans Electromagn Compat 51:499–518

Holle R, Murphy MJ (2015) Lightning in the North American monsoon: an exploratory climatology. Mon Weather Rev 143:1970–1977

Holle RL, Murphy MJ (2017) Lightning over three large tropical lakes and the strait of Malacca: exploratory analyses. Mon Weather Rev 145:4559–4573

Holle RL, Cummins KL, Brooks WA (2016) Seasonal, monthly, and weekly distributions of NLDN and GLD360 cloud-to-ground lightning. Mon Weather Rev 144:2855–2870

Holle RL, Said R, Scott M (2017) Global lightning variations. Meteor Tech Intl, Sept 129–140

Nag A, Murphy MJ, Schulz W et al (2015) Lightning locating systems: insights on characteristics and validation techniques. Earth Space Sci 2. https://doi.org/10.1002/2014EA000051

Nag A, Holle RL, Murphy MJ (2017) Cloud-to-ground lightning over the Indian subcontinent. In: Postprints of the 8th conference on the meteorological applications of lightning data, American Meteorological Society, Seattle, 22–26 Jan 2017

Orville RE (1991) Lightning ground flash density in the contiguous United States–1989. Mon Weather Rev 119:573–577

Orville RE, Huffines GR, Burrows WR et al (2011) The North American lightning detection network (NALDN)-analysis of flash data: 2001–09. Mon Weather Rev 139:1305–1322

Chapter 12
Time of Day and Time of Year of Lightning

Abstract Most lightning around the world occurs over land. On continents, the occurrence of lightning is controlled most often by heating of the land surface due to incoming solar radiation. As a result, there is a global afternoon maxima of lightning. In contrast, large-scale traveling meteorological disturbances can cause lightning to occur at any time of day. In addition, over the oceans away from land, there is much less variability in lightning through the day and night. During the course of the year, maximum lightning occurs during summer months in regions where middle latitude westerlies have distinct seasons. However, away from the westerlies, the annual cycle can be quite different due to movement of the equatorial trough and the Asian monsoon.

12.1 Time of Day

The diurnal pattern of US lightning is well known (Holle 2014). Most of the lightning occurs between 1200 and 1800 local time (Fig. 12.1). In general, two-thirds of lightning occur within these 6 afternoon hours when heating of the land is most intense and the highest temperatures of the day occur (Fig. 12.2).

The period from 1200 to 1800 local time typically has the most lightning due to the maximum daily temperature being reached after several hours of sunshine in the morning. Updrafts form, and the tops of the thunderstorms reach temperatures colder than freezing that are needed for lightning to occur (Chap. 13). After the storms have formed and matured, outflows and rainfall cool the surface, so that lightning production gradually diminishes in most situations. Lingering lightning may occur at the end of daytime storms and last into the early evening after sunset, but, in most locations, thunderstorms dissipate after dark.

The daily minimum in lightning is usually near 1000 local time. However, in some areas of the Central United States, there is a tendency for thunderstorms to persist through the night into the morning, particularly over the High Plains east of the Rocky Mountains, as shown in yellow to red in Fig. 12.2a. These storms originate in the afternoon but, due to complex meteorological conditions, sometimes propagate eastward across the plains to reach the Mississippi River and provide an

M. A. Cooper, R. L. Holle, *Reducing Lightning Injuries Worldwide*,
Springer Natural Hazards, https://doi.org/10.1007/978-3-319-77563-0_12

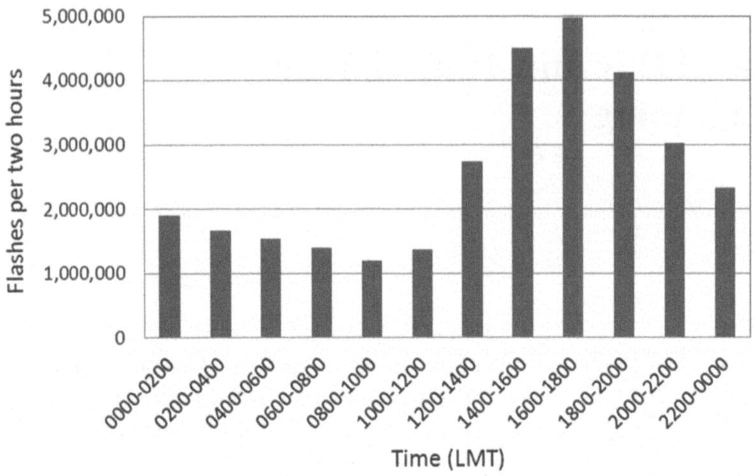

Fig. 12.1 Variations of cloud-to-ground flashes by local time of day over the contiguous United States and adjacent areas from the National Lightning Detection Network from 2005 through 2012. (Holle 2014)

early-morning maximum in some locations. This anomalous behavior has been found to be associated with high cloud bases and dry lower levels in the lee of large mountains in the middle latitudes and other aspects of a very different vertical charge structure in the thunderstorms than in most other locations (Carey and Buffalo 2007; Rust et al. 2005; Wiens et al. 2005).

Similar situations that lead to nighttime thunderstorms also exist in other locations of the world, but no studies similar to those in the United States have been pursued to date. This is an important issue since lightning can occur outside of the normally warmer afternoon hours and affect people and property in the midmorning. A recently identified midmorning persistence of lightning in Bangladesh is considered later in this section.

Questions to Explore

Does your region have frequent thunderstorms that persist into the night such as in the lee of the Rocky Mountains in the United States? If so, what are the meteorological conditions associated with these storms?

Globally, the afternoon maximum is dominant. Some locations around the world have the same conditions for storms to continue into the night as mentioned earlier over the Central United States. Another type of nighttime maximum is produced by a very different set of meteorological and topographic factors over the large tropical lakes of Lake Maracaibo in Venezuela, Lake Titicaca in western South America, and Lake Victoria in East Africa and the Strait of Malacca in Southeast Asia (Albrecht et al. 2016; Holle and Murphy 2017). These thunderstorms grow from the evening

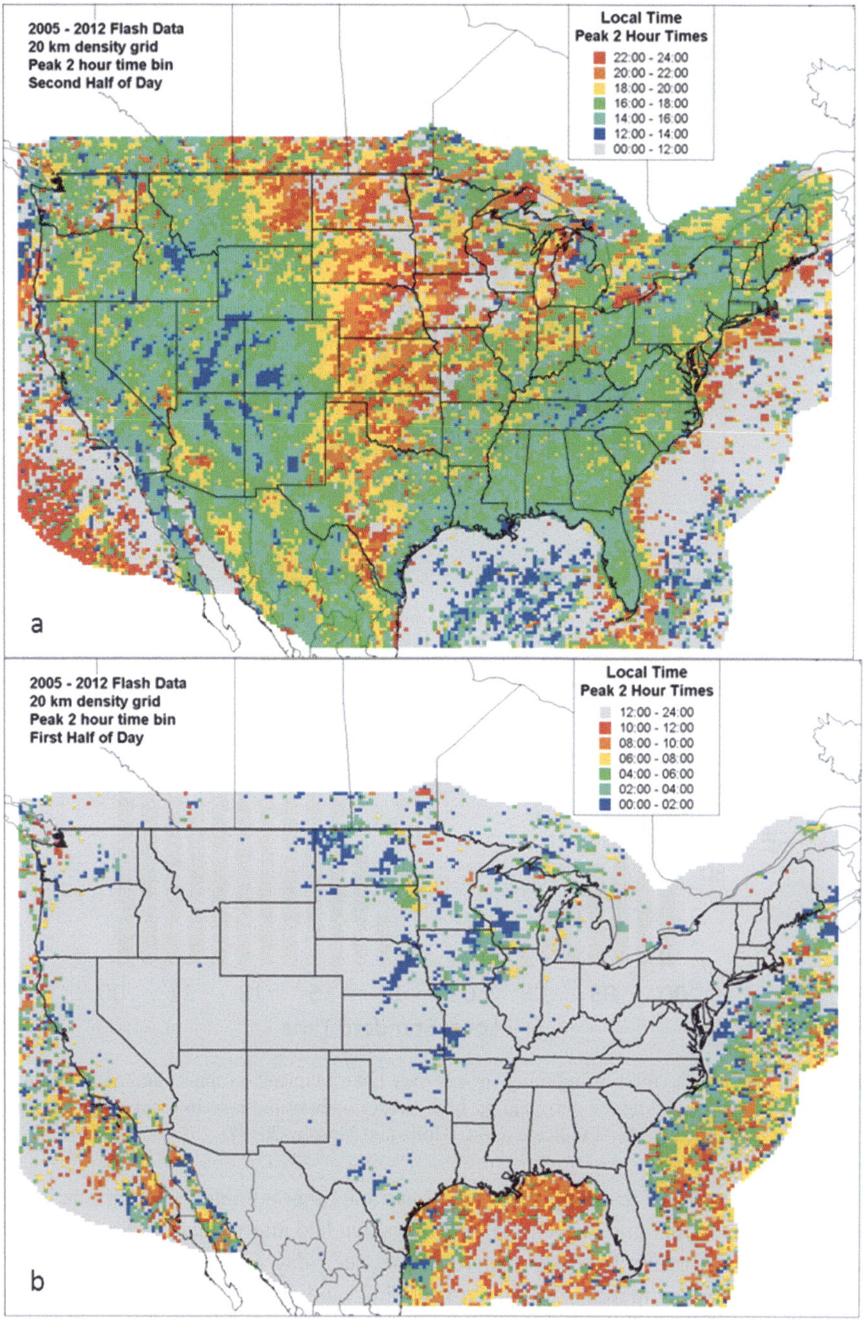

Fig. 12.2 Time of day of cloud-to-ground flashes over the contiguous United States and adjacent areas from the National Lightning Detection Network from 2005 through 2012. (Holle 2014)

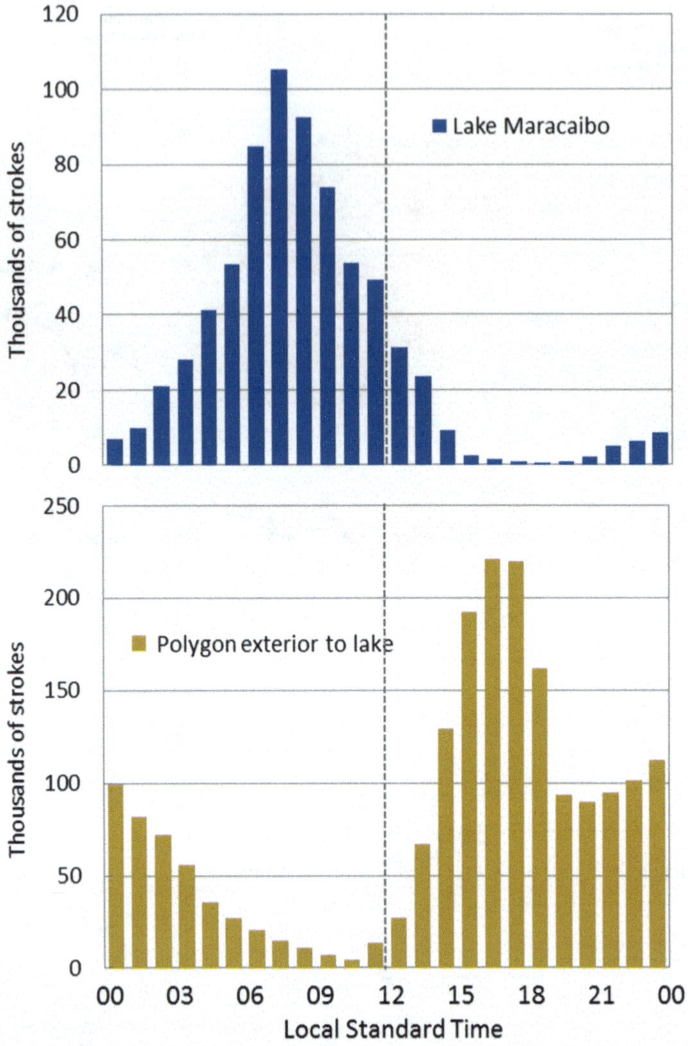

Fig. 12.3 Diurnal variation of lightning strokes over Lake Maracaibo (upper) and over a region within a polygon including the surrounding high terrain without the lake in local standard time (UTC-5). Dashed vertical lines indicate noon. (Holle and Murphy 2017)

through the middle of the night due to outflows and other complex interactions from adjacent large mountains surrounding the lakes that had afternoon thunderstorms on their slopes. An example of the difference between day and night lightning over Lake Maracaibo compared with the surrounding land areas is in Fig. 12.3. Such a nighttime occurrence may also prevail at some other tropical lakes surrounded by steep topography but have not yet been studied.

Because of the huge heat sink provided by large bodies of water, surface temperatures change very slowly over oceans compared with over land, so that the diurnal cycle for lightning is not observed over oceans. The formation of oceanic lightning is due to large-scale traveling upper lows, surface frontal boundaries, and other meteorological factors that are invariant through the daily cycle. Near land, however, oceanic lightning can be more frequent in the evening as daytime storms starting over land move over adjacent water but usually do not persist very long into the evening (Fig. 12.2).

Questions to Explore
Does your region have a large lake surrounded by steep topography that has nighttime lightning? How deep and how warm are these waters?

A recent study over the Indian subcontinent has depicted both the typical dominance of afternoon lightning and a notable exception. Figure 12.4 illustrates the

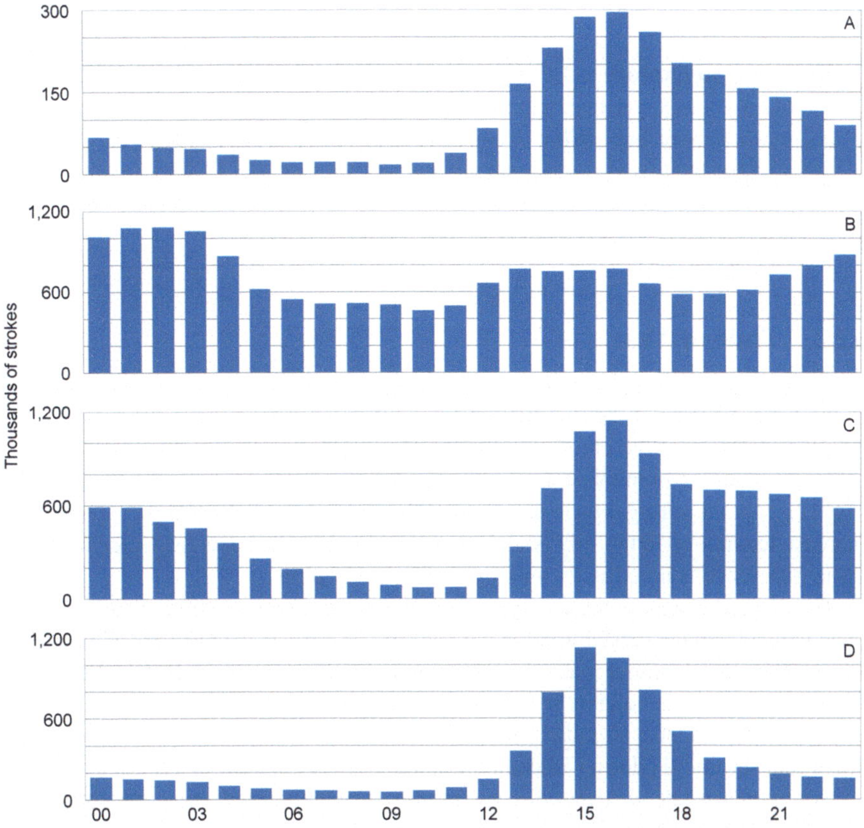

Fig. 12.4 Diurnal variation of strokes over the Indian subcontinent based on Global Lightning Dataset GLD360 data from 2012 to 2016. Region A is the broad northwest plains of India, B is the Bangladesh region, C is the southwest coast of India, and D is Sri Lanka. (Nag et al. 2017)

normal strong afternoon peak in strokes from noon to 1800 local time in three widely separated regions across the subcontinent. However, there is a midmorning persistence of lightning over Bangladesh that is similar to the afternoon frequency (Nag et al. 2017). This lack of a strong minimum from 0600 to.

1200 local time is not presently understood and has not been seen in most other locations around the world. Its existence is reinforced by the unusual situation in Bangladesh where the number of morning lightning fatalities is equal to those in the afternoon (Dewan et al. 2017).

Questions to Explore
Is this morning maximum in lightning over Bangladesh observed elsewhere and under what meteorological conditions? Are the scenarios of lightning deaths and injuries during this period different than those that occur in Bangladesh during the normal afternoon hours?

Questions to Explore
Based on your experience, when is the likely time of day when thunderstorms occur where you live? What exceptions from the expected time have sometimes happened? How could you validate your impressions? Do these impressions match the preceding results from various locations around the world?

12.2 Time of Year

In the United States, there is a distinct seasonal cycle of thunderstorms (Fig. 12.5). About two-thirds of the cloud-to-ground lightning flashes occur during the meteorological summer months of June, July, and August (Holle et al. 2016). Note in Fig. 12.5 that the spring season has more flashes than autumn. Similarly, the monthly increase leading up to the July peak is more gradual than the decrease after July; the same difference is noted in the monthly and weekly plots.

The United States pattern is typical of middle latitude locations where upper-level winds from the west dominate much of the year, such as in Europe, China, and Japan. The same summer maximum is apparent over the Southern Hemisphere land areas away from the equator at higher latitudes where westerlies dominate (generally poleward of 30° latitude). In these portions of South America, Africa, and Australia, summer is during December, January, and February.

Continental-scale distributions by month throughout the year are shown in Fig. 12.6. The data are shown by percentages relative to the annual total for each continent. The Northern Hemisphere typically has the most lightning in July and much smaller percentages in winter. Lightning occurrence over the Southern

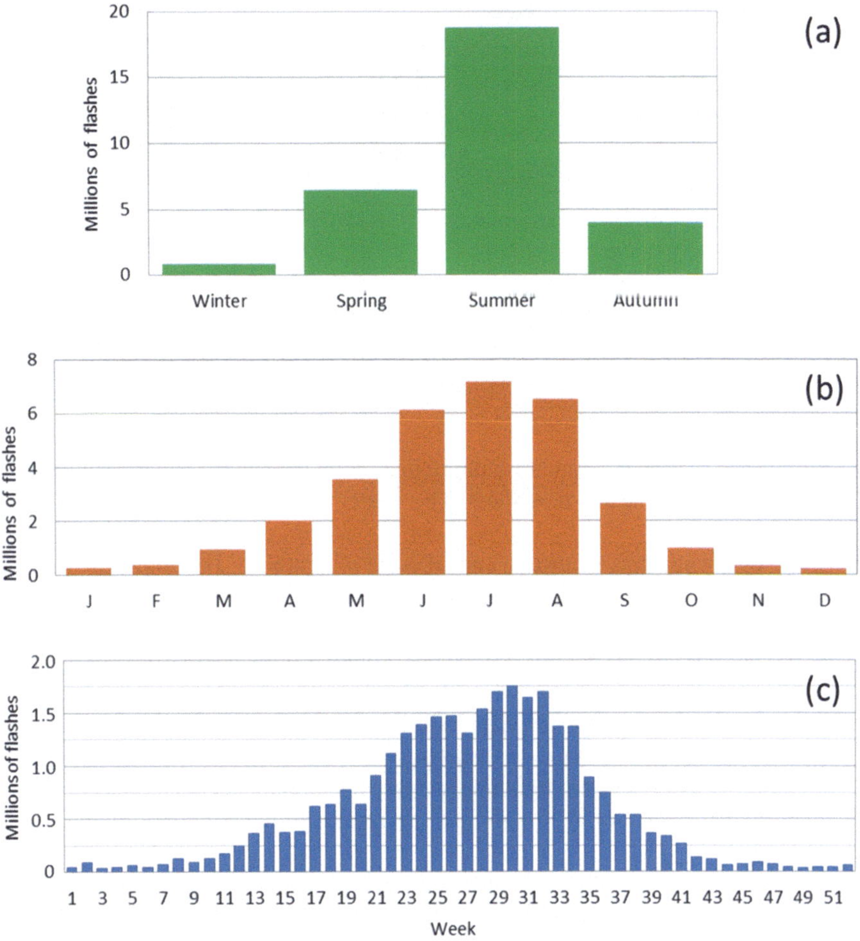

Fig. 12.5 Variations of cloud-to-ground flashes by (**a**) season, (**b**) month, and (**c**) week over the contiguous United States and adjacent areas based on 310,162,364 cloud-to-ground flashes from the National Lightning Detection Network from 2005 through 2014. (Holle et al. 2016)

Hemisphere continents is reversed and has peak lightning activity in the opposite months (Fig. 12.4). African lightning frequency is close to being uniform through the year due to its location on both sides of the equator.

The following relates to each continent in Fig. 12.6, beginning in the upper left panel:

- *North America* starts with minimal lightning in the first few months of the year, followed by a steady increase to the July peak, then a reduction into the autumn months.
- *South America* has an annual lightning cycle that is nearly opposite to the Northern Hemisphere. Lightning is most frequent near the first of the year and is at a minimum during June and July when drier low-level air prevails over all except the Amazon basin, where strokes persist all year.

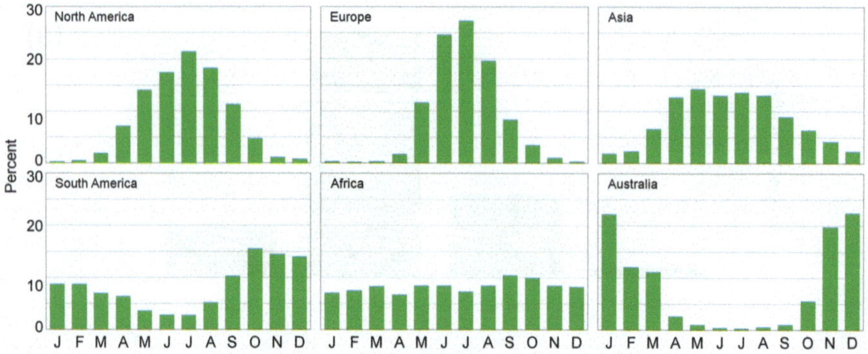

Fig. 12.6 Lightning stroke percentages by month for each continent. (Holle et al. 2017)

- *Europe* has a sharper summer lightning maximum than North America, although the number of strokes is much less. The winter period has a very small amount of lightning.
- *Africa* as a whole has almost no annual cycle in lightning. The continent's equatorial regions have activity throughout the year that depends on the location of the equatorial trough during the year. Away from the equator (generally poleward of 30° latitude), the Northern Hemisphere summer influences stroke occurrence in Africa north of the equator, while the Southern Hemisphere summer results in frequent lightning to the south. The resulting annual cycle for the entire continent in Fig. 12.6 shows a nearly even distribution through the year. However, there are strong monthly changes in lightning at every location that depend on its position relative to the equatorial trough and the Northern and Southern Hemisphere summers.
- *Asia* is mostly north of the equator and has higher percentages of strokes from May through August. However, the equatorial regions of Asia have strokes all year so that winter months have larger percentages of lightning than over North America and Europe.
- *Australia* has a well-defined Southern Hemisphere annual cycle of lightning since the entire continent is south of the equator. Monthly stroke percentages peak around the first of the year and are quite small in winter.
- *Antarctica* has no detected lightning.

In tropical regions where winds generally have an easterly component, the time of year of maximum lightning is dominated by the passage of the equatorial trough, also called the intertropical convergence zone. Over Africa in the Northern Hemisphere summer, the trough moves as far north as the southern border of the Sahara. In winter, the equatorial trough moves southward to south of the equator. As a result, some regions near the equator have two lightning peaks as the equatorial trough crosses overhead twice during the year, while locations poleward of the equator have one dominant thunderstorm season. Thunderstorms can occur all year over and near the equator, in such regions as Africa, South America, and Southeast Asia so that large lightning frequencies are apparent in Fig. 11.2 in these locations.

A different annual cycle applies to the large region dominated by the Asian monsoon. Lightning over the Indian subcontinent has begun to be explored by Nag et al. (2017). During the winter months of November through February, northerly flow from the Asian continent pours dry air southward and dominates locations from India through southern China, so that thunderstorms are infrequent. With the arrival of the Northern Hemisphere spring, deep tropical moisture starts to flow back northward and results in a region of thunderstorms that hovers across Southeast Asia for several months (Nag et al. 2017).

Over the oceans, the annual cycle of lightning is muted. Instead, meteorological factors dominate lightning production. Cold-core low-pressure systems aloft occasionally produce deep vertical instability that results in lightning over oceans quite far from the equator in both the Northern and Southern Hemispheres (Figs. 11.2 and 11.3). It should also be pointed out that lightning over oceans is more frequent closer to land, compared with far from land, due to thunderstorms that originate over continents and large islands and move downstream depending on the prevailing meteorological flow in each region. These downwind lightning maxima are especially evident off the east coasts of all three Southern Hemisphere continents at higher latitudes (Figs. 11.2 and 11.3).

Questions to Explore

Based on your experience, when is the likely time of year when thunderstorms occur where you live? Is there more than one time during the course of the year? Do these impressions match with the preceding results from various locations around the world? Where could you search for data to validate your experiences and impressions?

References

Albrecht R, Goodman SJ, Buechler DE et al (2016) Where are the lightning hotspots on earth? Bull Am Meteorol Soc 97:2051–2068

Carey LD, Buffalo KM (2007) Environmental control of cloud-to-ground lightning polarity in severe storms. Mon Weather Rev 135:1327–1353

Dewan A, Hossain MMF, Rahman MM et al (2017) Recent lightning-related fatalities and injuries in Bangladesh. Weather Clim Soc 9:575–589

Holle RL (2014) Diurnal variations of NLDN-reported cloud-to-ground lightning in the United States. Mon Weather Rev 142:1037–1052

Holle RL, Murphy MJ (2017) Lightning over three large tropical lakes and the strait of Malacca: exploratory analyses. Mon Weather Rev 145:4559–4573

Holle RL, Cummins KL, Brooks WA (2016) Seasonal, monthly, and weekly distributions of NLDN and GLD360 cloud-to-ground lightning. Mon Weather Rev 144:2855–2870

Holle RL, Said R, Scott M (2017) Global lightning variations. Meteor Tech Intl, Sept 129–140

Nag A, Holle RL, Murphy M (2017) Cloud-to-ground lightning over the Indian subcontinent. In: Postprints of the 8th conference on the meteorological applications of lightning data. American Meteorological Society, Seattle/Washington. 22–26 Jan 2017

Rust WD, MacGorman DM, Bruning EC et al (2005) Inverted-polarity electrical structures in thunderstorms in the severe thunderstorm electrification and precipitation study (STEPS). Atmos Res 76:247–271

Wiens KC, Rutledge SA, Tessendorf SA (2005) The 29 June 2000 supercell observed during STEPS. Part II: lightning and charge structure. J Atmos Sci 62:4151–4177

Chapter 13
Meteorological Concepts Affecting Lightning Formation

Abstract Lightning always has its origin in clouds whose tops are colder than freezing, and updrafts connect the lower portions of a thunderstorm with an altitude where clouds are colder than freezing. The freezing level during the warm months of the year in middle latitude and tropical regions is often between 4.2 and 5.2 km (14,000 and 17,000 feet) but can be much closer to the ground in winter or in colder regions. Lightning initiates at temperatures between −5 and −15 °C (23 and 5 F), so the layer where lightning forms is a kilometer (3000 feet) or more higher in altitude than the freezing level.

13.1 Updrafts

The atmosphere is constantly in motion and does not flow horizontally. Large-scale traveling meteorological systems during winter in North and South America, Europe, Northern Asia, and parts of Africa and Australia more than about 30° latitude away from the equator can have horizontal winds of 50 meters per second (110 miles per hour/177 km per hour) or more. They have much weaker vertical motions, on the order of centimeters per second (less than one mile per hour), but they persist for many hours and lift air upward far enough to produce clouds and precipitation over broad areas.

In contrast, upward motions that result in lightning are called updrafts within cumulus clouds. The updraft can be 5 meters per second (11 miles per hour) and may reach 25 meters per second (56 miles per hour) or more. These strong upward motions are concentrated in well-defined columns of air whose cores are typically at least 3 km (2 miles) or more in diameter. While these towers occupy a small amount of time and space, they are the essential ingredients of cloud-to-ground lightning production (Holle 1984). Figure 13.1 shows examples of updraft towers within cumulus clouds; cloud types are described more completely in Sect. 13.3. Note that in-cloud lightning is more frequent than cloud-to-ground lightning and typically results from less intense upward motions.

An updraft resulting in lightning starts at altitudes where the air is warmer than freezing, and then upward motion connects the warmer lower levels with cloudy

M. A. Cooper, R. L. Holle, *Reducing Lightning Injuries Worldwide*,
Springer Natural Hazards, https://doi.org/10.1007/978-3-319-77563-0_13

Fig. 13.1 Photograph of numerous cumulus towers containing updrafts indicated by red arrows over Colorado. (©R. Holle)

areas that are colder than freezing. The necessary ingredients for lightning production are known to be an updraft and a mixture of small hail called graupel, supercooled water droplets, and ice particles (Takahashi 1978; Saunders 1993; Stolzenburg et al. 1998; Williams 2001; Nag and Rakov 2012; Rakov 2016). The larger particles leave the updraft when they become too heavy to be kept aloft or are thrown outward from the updraft.

Meanwhile, the lighter-weight particles stay aloft. The difference in size of ice and water particles results in separation of charge and initiates lightning. Heavier particles, such as hail, fall more quickly than light ice crystals and carry charge with them to lower altitudes. At that time, transfer of charge between colliding particles assists in the development of cloud-scale layers with differing charge (Williams 2001). Exact details of how lightning starts are not well known, and the relative influence of these factors varies somewhat at different times and locations (MacGorman and Rust 1998).

The driving force for lightning production is an updraft that organizes the vertical motions, regardless of the underlying meteorological conditions, listed in Sect. 13.2. The tops of updrafts are visible in Fig. 13.1 by the rounded tops, which represent buoyant bubbles of air moving directly upward. There are numerous updraft towers in the photo, and the tallest and largest are capable of producing lightning. Other updraft towers in the lower portion of the photo may later become tall enough to reach altitudes with temperatures colder than freezing and result in lightning.

These vertical motions result from a large temperature difference in the vertical, where the warmest air is at the lower levels and colder air is aloft. Small parcels of air warmer than the surrounding atmosphere begin to rise in a column due to a variety of causes listed in Sect. 13.2. Instability develops when the temperature inside the buoyant bubble is warmer than in the surrounding atmosphere. Depending on the vertical temperature structure in the atmosphere outside of the cloud, the buoyant bubble may start to accelerate as it rises since it becomes still warmer than the surrounding air. That difference can increase with height until the buoyant bubble is much warmer than the surrounding air, and it moves upward still more quickly. When updrafts carry particles from lower altitudes to the subfreezing layer of −5 to −15 °C (23–5 F), separation of charge initiates lightning between the various sizes and types of frozen and liquid particles.

Eventually, the tower takes on the same temperature as the surrounding atmosphere and the buoyant bubble stops rising. In Fig. 13.1, the tallest towers continue to have hard outlines on top, which represent updrafts. All of the clouds in the photo have rounded tops, which indicate that they are warmer than the surrounding air at this time. The veiled horizontal sheets (pileus) on some of the towers indicate locations where the updrafts are penetrating a layer of the atmosphere that is moister than other layers, so the rising air pushes the moist layer upward until saturation is reached.

13.2 Meteorological Conditions Resulting in Lightning

The cores of the updrafts within cumulus towers are similar wherever they form. However, the surrounding atmosphere that causes the updraft tower can result from widely varying conditions, including the following types:

- *Turbulence* in the atmosphere is always present in varying intensities. Although land may be completely flat, there is some turbulent motion that can result in upward motion. Over strongly heated flatland, some of the upward motions become strong enough during the middle of the day to coalesce into updrafts that result in lightning-producing clouds. The locations of such thunderstorms are essentially random and not caused by local features. Such storms are likely to have only a few updraft towers, be less than 10 km (6 miles across), and last only an hour or so. The lightning-producing phase of such a storm may only last tens of minutes. These short-lived thunderstorms are common over the broad flat landmasses of equatorial regions and land areas in summer months further from the equator.
- *Sea breezes* initiate updrafts near coastlines of oceans and other very large bodies of water. Each day, the direction and velocity of the winds throughout the entire depth of the atmosphere affect which coastlines are affected and how intense the activity will become later in the day. The temperature and dew point, throughout the depth of the atmosphere, affect the intensity of the sea breeze. The timing, location, and intensity of the sea breezes over the state of Florida in the United States have been studied extensively with both observations and modeling (Pielke 1974). With regard to lightning, subsequent research has taken into account these factors and resulted in substantial skill in anticipating how lightning from the day's thunderstorms is likely to develop over the classic sea breeze situation of the Florida Peninsula (Lericos et al. 2002; Shafer and Fuelberg 2006; Rudlosky and Fuelberg 2011). Similar sea breeze studies have taken place along the United

States Gulf of Mexico coast (Camp et al. 1998; Smith et al. 2005). Such approaches have been especially important for the rocket launch facilities at Kennedy Space Center, located on the east coast of Florida (Lambert et al. 2006).

- *Major elevation changes* affect where updrafts are formed by air that is forced to change altitude and lifted vertically due to very large variations in the underlying terrain. Daily and hourly changes in lightning over the state of Colorado in the United States have indicated that the most frequent lightning tends to occur on the slopes of steep terrain and not over the highest peaks (Cummins 2012; Vogt and Hodanish 2014, 2016). The result is that storms forming first over the highest peaks often produce few cloud-to-ground flashes that are, unfortunately, not perceived at the time to be a major threat to people (Hodanish et al. 2004, 2015). Additional locations have been examined around the world and in tropical regions where it has also been found that lightning occurs most often on the slopes of high peaks and mountain ranges (Holle and Murphy 2015, 2017). The maxima face toward the direction of the strongest low-level moist flow arriving at a location (Cummins 2012). Small local hills, such as those that are tens of meters high, are not nearly large enough to force an updraft strong enough to reach the subfreezing layer where lightning is formed.
- *Severe thunderstorms* are defined by the Storm Prediction Center (SPC) of the US National Weather Service (www.spc.noaa.gov) as a thunderstorm producing hail that is at least 1 inch (2.5 cm) in diameter or larger, and/or wind gusts of 58 mile/hour (93 km per hour) or greater, and/or a tornado. A large number of studies have focused on the highly variable lightning features associated with severe weather, such as polarity reversal of cloud-to-ground flashes, differing vertical structures of charge layers, variations in cloud-to-ground versus in-cloud lightning, organizational structure, and the accompanying radar depictions in multiple locations around the globe (Keighton et al. 1991; Soula et al. 2004; Carey and Buffalo 2007; Schultz et al. 2007; Xu et al. 2016; Zheng and MacGorman 2016). These variations are often geographically dependent and do not result in a consistent occurrence of the same severe weather at any given location. A specific project took place to investigate the meteorological factors affecting lightning production associated with severe weather over the US High Plains (Lang et al. 2004). Weekly severe weather probabilities from SPC are compared with lightning frequencies over the United States by Holle et al. (2016).
- *Mesoscale convective systems* (MCSs) are prolific lightning producers (Steiger et al. 2007; Dotzek et al. 2005). MCSs cover very large areas, as much as 100,000 square kilometers or more in some cases. They tend to occur over land, are strongest at night, last for up to 18 h, and have been measured to produce tens of thousands of cloud-to-ground flashes in a single night (Laing and Fritsch 1997). They tend to occur downwind of large terrain gradients on the subcontinental scale such as the Central United States and Argentina, among other locations. The longest lightning events measured in time and distance have both been associated with MCSs as described in Chap. 13.3 (Lang et al. 2017). The updraft is not so much of a pure vertical column in such mesoscale systems but develops into a large sloping layer that can persist in a steady state for many hours if the

environmental conditions are in the correct positions (Ely et al. 2008). Severe weather may accompany an MCS, more often in the earlier portion of its lifecycle.

- *Derechos* are the source of widespread strong winds in the evening and nighttime, primarily in the summer in middle latitudes (Johns and Hirt 1987; Bentley and Mote 1998). They may produce prolific lightning along their path of rapid movement from a westerly component in the Northern Hemisphere.
- *Large-scale systems* without significant organization of severe weather may also produce widespread but less intense lightning (van den Broeke et al. 2005). Sufficiently strong updrafts to produce lightning can be found in such meteorological features as cold fronts, squall lines, low-pressure troughs, and beneath upper-level lows. Their sources of meteorological development are similar to the severe weather situations described above but are not intense enough to produce high winds, tornadoes, or hail.
- *Tropical cyclones* can produce lightning in some situations (Stevenson et al. 2016). A consistent profile of lightning within tropical cyclones has not emerged to date with respect to growth or weakening (DeMaria et al. 2012; Zhang et al. 2015). The outer rainbands tend to have more lightning than the eyewall in most storms, but exceptions exist. When the outer rainbands come onshore, regardless of the intensity of the cyclone, frequent lightning may occur over heated landmasses, and there may be associated heavy rainfall (Molinari et al. 1999). Structural features such as shear have been explored, but no single parameter has proved to be a clear indication of storm phase (Molinari et al. 2004; Wang et al. 2016). Tropical storms tend to have more lightning than hurricane force systems. Future research with larger databases is indicated.
- *Rainfall amount* is not always directly related to lightning frequency. Correlations of lightning with precipitation have been made on scales ranging from individual storms up to monthly periods using comparisons with rain gages, radar, and satellite data around the world (Chèze and Sauvageot 1997; Kar and Ha 2003; Kempf and Krider 2003; Petrova et al. 2014; Zheng et al. 2010). Very large rates of lightning usually indicate heavy rain at the ground. However, there can be heavy rain with little or no lightning in tropical regions when the atmosphere is very moist and has weak vertical instability. A threat for forest fire ignition in more arid regions is dry lightning, which results from high-based thunderstorms that are above dry lower levels of the atmosphere that are not accompanied by enough rainfall to extinguish a forest fire (Nauslar et al. 2013).
- *Winter storms* occasionally have lightning in three scenarios (March et al. 2016). One situation is lake-effect snowbands downwind of the US Great Lakes (Steiger et al. 2009) and sea-effect storms over western Japan (Wu et al. 2014). While the parent thunderstorms are very shallow, these lightning events have unusual features that make them very damaging to electrical and wind turbine installations. These storms only occur when very cold air flows across a large body of water that is not frozen. The second situation occurs when large-scale processes over land far from the ocean result in instability aloft above an air mass near the ground that is below freezing (Market and Becker 2009). The third situation is when large low-pressure systems produce snow and have embedded lightning

(Rauber et al. 2014). Casualties are rare in winter storms due to a combination of small flash rates and few people being outside during such storms.

Questions to Explore

What types of meteorological conditions most commonly produce lightning where you live? Do other regions of your country tend to have a different set of conditions? At what time of day and year are they likely to be a threat to people and property in each of these regions?

13.3 Characteristics of Lightning

Several features of the processes involved in lightning production are shown in Fig. 13.2. Lightning is initiated when a cumulonimbus cloud has penetrated the layer between about −5 and −15 °C (23 and 5 F) in a buoyant bubble containing an updraft. Cloud-to-ground flashes initiated in this subfreezing layer have downward branching as in Fig. 13.2. The bright primary channels are mostly vertical, and the downward direction is apparent in the unsuccessful angled branches toward ground on both sides of the two channels. The downward movement of a cloud-to-ground flash is made in steps that are about 50 m (160 feet) long as indicated by the jagged shape of the lightning channels in Fig. 13.2. When a downward leader comes within 30–50 m (100–160 feet) of the ground, it searches to make a connection with an object on the surface

Fig. 13.2 Photograph of two cloud-to-ground lightning flashes over Arizona. (©R. Holle)

of the earth. In a sense, the lowest tip of the downward-propagating leader can be visualized as a pendulum while it searches for the closest connection with an object on the ground. When that connection is made, the bright visible light travels up the existing faintly visible downward-propagating channel. At that time, the unsuccessful side branches disappear. This process occurs at about one-third the speed of light.

Questions to Explore
Is it likely that a person can see the faint portion of a cloud-to-ground lightning flash coming to ground or separate the downward- from upward-moving path? Is there enough time to react to a nearby lightning flash in order to reach a safe vehicle or building?

A cloud-to-ground flash has one or more return strokes; the average is four to five cloud-to-ground strokes per cloud-to-ground flash (Rakov 2016). When lightning appears to be flickering, those are individual return strokes within the flash. Return strokes are somewhat weaker than the original stroke, are less than a tenth of a second apart, and usually follow the same channel as the original stroke in a cloud-to-ground flash (Nag et al. 2008). However, in about half of cloud-to-ground flashes, one of the subsequent strokes branches away from the main channel and comes to ground within 2–3 km (1–2 miles) of the pre-existing channel that the previous strokes had followed. Occasionally a flash can reach outward up to 25 km (16 miles) from the channel (Rison et al. 2003). Figure 13.3 shows a multiple-stroke flash that was obtained by

Fig. 13.3 Photograph of a multi-stroke cloud-to-ground flash over Arizona. (©R. Holle)

Fig. 13.4 Photograph of a long in-cloud flash at night over Oklahoma. (©R. Holle)

moving the camera from right to left while the lightning occurred. Note that downward branching apparent in Fig. 13.2 is only emanating from the first stroke located on the right side of the photo. About 90% of cloud-to-ground flashes lower negative charge; the rest lowers positive charge to ground (Nag et al. 2015).

In the majority of storms, there are three to four times as many in-cloud flashes as cloud-to-ground flashes. Figure 13.4 shows an in-cloud lightning flash that is horizontally extensive. Cloud lightning flashes do not directly result in damage or casualties on the ground, but they must be monitored for safety along with cloud-to-ground flashes. In-cloud lightning indicates that processes are occurring that may result in lightning reaching the ground at the same time and place as the in-cloud flashes (Chap. 14). Such in-cloud flashes can become very extensive in MCSs. Lang et al. (2017) document a continuous in-cloud flash that traveled 321 km (199 miles) without interruption over Oklahoma in the United States and another that lasted 7.74 s over France. Chapter 14 describes how these events were detected.

13.4 Cloud Types

Context for the types of clouds that produce lightning is now illustrated by a sequence of how cumulus clouds form (Holle 2014). Cumulus clouds have shapes whose upper boundaries appear in puffs, mounds, or towers that have a vertical or somewhat slanted appearance. The other broad type of cloud is stratiform that is generally horizontal (Ludlum et al. 1995).

Fig. 13.5 Photograph of cumulus humulis clouds over Barbados. (©R. Holle)

Cumulus clouds in the early flat stage are called cumulus humulis – they are wider than they are tall (Fig. 13.5). After they become taller than wide, the result is a cumulus congestus cloud (Fig. 13.6). Finally, a cumuliform cloud that continues to grow may become a cumulonimbus with flattened tops spreading outward (Fig. 13.7).

Questions to Explore
Is the sky usually clear enough for you to see these cloud features or is there haze and pollution blocking your view most of the time? How often do intervening stratiform clouds block the visibility of updraft towers?

For Further Exploration
The World Meteorological Organization of the United Nations has recently published a thorough summary and classification of all cloud types at https://www.wmocloudatlas.org/home.html. This cloud atlas includes photos of the clouds, their descriptions, the associated upper air soundings, and the surface weather conditions and points out similarities and differences with other cloud types. Some clouds and phenomena are unique to specific locations in the world, so that a person will never see all of them without a significant amount of travel!

Fig. 13.6 Photographs of a single cumulus congestus cloud over Arizona on top, and row of congestus over Oklahoma on bottom. (©R. Holle)

Fig. 13.7 Photographs of a single cumulonimbus cloud on top, and complex of cumulonimbi over Arizona on bottom. (©R. Holle)

References

Bentley JL, Mote TL (1998) A climatology of derecho-producing mesoscale convective systems in the central and eastern United States, 1986–1995. Part I: temporal and spatial distribution. Bull Am Meteorol Soc 79:2527–2540

Carey LD, Buffalo KM (2007) Environmental control of cloud-to-ground lightning polarity in severe storms. Mon Weather Rev 135:1327–1353

Chèze J-L, Sauvageot H (1997) Area-average rainfall and lightning activity. J Geophys Res 102:1707–1715

Cummins KL (2012) Analysis of multiple ground contacts in cloud-to-ground flashes using LLS data: the impact of complex terrain. In: Preprints of the 22nd international lightning detection conference, Broomfield, 2–3 Apr 2012

Camp JP, Watson AI, Fuelberg HE (1998) The diurnal distribution of lightning over North Florida and its relation to the prevailing low-level flow. Weather Forecast 13:729–739

DeMaria M, DeMaria TT, Knaff JA et al (2012) Tropical cyclone lightning and rapid intensity change. Mon Weather Rev 140:1828–1842

Dotzek N, Rabin RM, Carey LD et al (2005) Lightning activity related to satellite and radar observations of a mesoscale convective system over Texas on 7-8 April 2002. Atmos Res 76:127–166

Ely BL, Orville RE, Carey LD et al (2008) Evolution of the total lightning structure of a leading-line, trailing-stratiform mesoscale convective system over Houston, Texas. J Geophys Res 113:D08114. https://doi.org/10.1029/2007JD008445

Hodanish S, Holle RL, Lindsey DT (2004) A small updraft producing a fatal lightning flash. Weather Forecast 19:627–632

Hodanish S, Wolyn P, Mozley K (2015) Meteorological analysis of the Rocky Mountain National Park lightning fatalities of 11 and 12 July, 2014. In: Preprints of the 7th conference on the meteorological applications of lightning data, American Meteorological Society, Phoenix, 4–8 Jan 2015

Holle RL (1984) Spherical cumulonimbus tower. Bull Am Meteorol Soc 65:371–372

Holle RL (2014) The visual identification of lightning-producing thunderstorm clouds. In: Preprints of the 5th international lightning meteorology conference, Vaisala, Tucson, 20–21 Mar 2014

Holle R, Murphy MJ (2015) Lightning in the north American monsoon: an exploratory climatology. Mon Weather Rev 143:1970–1977

Holle RL, Murphy MJ (2017) Lightning over three large tropical lakes and the strait of Malacca: exploratory analyses. Mon Weather Rev 145:4559–4573

Holle RL, Cummins KL, Brooks WA (2016) Seasonal, monthly, and weekly distributions of NLDN and GLD360 cloud-to-ground lightning. Mon Weather Rev 144:2855–2870

Johns RH, Hirt WD (1987) Derechos: widespread convectively induced windstorms. Weather Forecast 2:32–49

Kar SK, Ha K-J (2003) Characteristic differences of rainfall and cloud-to-ground lightning activity over South Korea during the summer monsoon season. Mon Weather Rev 131:2312–2323

Keighton SJ, Bluestein HB, MacGorman DR (1991) The evolution of a severe mesoscale convective system: cloud-to-ground lightning location and storm structure. Mon Weather Rev 119:1533–1556

Kempf NM, Krider EP (2003) Cloud-to-ground lightning and surface rainfall during the great flood of 1993. Mon Weather Rev 13:1140–1149

Laing AG, Fritsch JM (1997) The global population of mesoscale convective complexes. Q J Roy Meteorol Soc 123:389–405

Lambert W, Wheeler M, Roeder W (2006) Lightning probability forecasts at Kennedy Space Center/Cape Canaveral air Force Station, Florida. In: Preprints of the 1st international lightning meteorology conference, Vaisala, Tucson, 26–27 Apr 2006

Lang TJ, Miller LJ, Weisman M et al (2004) The severe thunderstorm electrification and precipitation study (STEPS). Bull Am Meteorol Soc 85:1107–1125

Lang TJ, Pédeboy S, Rison W et al (2017) WMO world record lightning extremes: longest reported flash distance and longest reported flash duration. Bull Am Meteorol Soc 98:1153–1168

Lericos TP, Fuelberg HE, Watson AI et al (2002) Warm season lightning distributions over the Florida peninsula as related to synoptic patterns. Weather Forecast 17:83–98

Ludlum DM, Holle RL Keen RA (1995) The Audubon Society pocket guide: clouds and storms. Alfred A. Knopf

MacGorman D, Rust WD (1998) The electrical nature of storms. Oxford Univ Press, New York, New York, U.S.

Market P, Becker AE (2009) A study of lightning flashes attending periods of banded snowfall. Geophys Res Lett. https://doi.org/10.1029/2008GL036317

Molinari J, Moore P, Idone VP (1999) Convective structure of hurricanes as revealed by lightning location. Mon Weather Rev 127:520–534

Molinari J, Vollaro D, Corbosiero KL (2004) Tropical cyclone formation in a sheared environment: a case study. J Atmos Sci 61:2493–2509

March V, Montanyà J, Fabro F et al (2016) Winter lightning activity in specific global regions and implications to wind turbines and tall structures. In: Preprints of the 33rd international conference on lightning protection, Estoril, 25–30 Sept 2016

Nag A, Rakov VA (2012) Positive lightning: an overview, new observations, and inferences. J Geophys Res 117. https://doi.org/10.1029/2012JD017545

Nag A, Murphy MJ, Schulz W et al (2015) Lightning locating systems: insights on characteristics and validation techniques. Earth Space Sci 2. https://doi.org/10.1002/2014EA000051

Nag A, Rakov VA, Schulz W et al (2008) First versus subsequent return-stroke current and field peaks in negative cloud-to-ground lightning discharges. J Geophys Res 113:D19112. https://doi.org/10.1029/2007JD009729

Nauslar NJ, Kaplan ML, Wallmann J et al (2013) A forecast procedure for dry thunderstorms. J Oper Meteorol 1:200–214

Petrova S, Mitzeva R, Kotroni V (2014) Summer-time lightning activity and its relation with precipitation: diurnal variation over maritime, coastal and continental areas. Atmos Res 135–136:388–396

Pielke RA (1974) A three-dimensional numerical model of the sea breezes over South Florida. Mon Weather Rev 102:115–139

Rakov VA (2016) Fundamentals of lightning. Cambridge Univ Press, Cambridge, U.K.

Rauber RM, Wegman J, Plummer DM et al (2014) Stability and charging characteristics of the comma head region of continental winter cyclones. J Atmos Sci 71:1559–1582

Rison W, Krehbiel P, Thomas R et al (2003) Lightning mapping and radar observations of bolts from the blue. In: Proceedings of the 12th international conference on atmospheric electricity, Versailles, 9–13 June 2003

Rudlosky SD, Fuelberg HE (2011) Seasonal, regional, and storms-scale variability of cloud-to-ground lightning characteristics in Florida. Mon Weather Rev 139:1826–1843

Saunders CPR (1993) A review of thunderstorm electrification processes. J Appl Meteorol 32:642–655

Schultz DM, Weiss CC, Hoffman M (2007) The synoptic regulation of dryline intensity. Mon Weather Rev 135:1699–1709

Shafer PE, Fuelberg HE (2006) A statistical procedure to forecast warm season lightning over portions of the Florida peninsula. Weather Forecast 21:851–868

Smith JR, Fuelberg HE, Watson AI (2005) Warm season lightning distributions over the northern Gulf of Mexico coast and their relation to synoptic-scale and mesoscale environments. Weather Forecast 20:415–438

Soula S, Sauvageot H, Molinié GF et al (2004) Cloud-to-ground lightning activity in hail-bearing storms. J Geophys Res. https://doi.org/10.1029/2003JD003669

Steiger SM, Orville RE, Carey LD (2007) Total lightning signatures of thunderstorm intensity over North Texas. Part II: Mesoscale convective systems. Mon Weather Rev 135:3303–3324

Steiger SM, Hamilton R, Keeler J et al (2009) Lake-effect thunderstorms in the lower Great Lakes. J Appl Meteorol Climatol 48:889–902
Stevenson SN, Corbosiero KL, Abarca SF (2016) Lightning in eastern North Pacific tropical cyclones: a comparison to the North Atlantic. Mon Weather Rev 144:225–239
Stolzenburg M, Rust WD, Marshall TC (1998) Electrical structure in thunderstorm convective regions. Part 3: synthesis. J Geophys Res 103(D12):14097–11410
Takahashi T (1978) Riming electrification as a charge generation mechanism in thunderstorms. J Atmos Sci 35:1536–1548
van den Broeke MS, Schultz DM, Johns RH et al (2005) Cloud-to-ground lightning production in strongly forced, low-instability convective lines associated with damaging wind. Weather Anal Forecast 20:517–530
Vogt BJ, Hodanish SJ (2014) A high-resolution lightning map of the state of Colorado. Mon Weather Rev 142:2353–2360
Vogt BJ, Hodanish SJ (2016) A geographical analysis of warm season lightning/landscape interactions across Colorado, USA. Appl Geogr 75:93–103
Wang F, Qie X, Liu D et al (2016) Lightning activity and its relationship with typhoon intensity and vertical wind shear for super typhoon Haiyan (1330). J Meteorol Res 30:117–127
Williams E (2001) The electrification of severe storms. In: Doswell (ed) severe convective storms. Am Meteorol Soc, Boston, pp 527–261
Wu T, Yoshida S, Ushio T et al (2014) Large bipolar lightning discharge events in winter thunderstorms in Japan. J Geophys Res Atmos 119:555–566
Xu S, Zheng D, Wang Y et al (2016) Characteristics of the two active stages of lightning activity in two hailstorms. J Meteorol Res 30:265–281
Zhang W, Zhang Y, Zheng D et al (2015) Relationship between lightning activity and tropical cyclone intensity over the Northwest Pacific. J Geophys Res Atmos 120:4072–4089
Zheng D, MacGorman DR (2016) Characteristics of flash initiations in a supercell cluster with tornadoes. Atmos Res 167:249–264
Zheng DY, Zhang J, Meng Q et al (2010) Relationship between lightning activities and surface precipitation in thunderstorm weather in Beijing. Chinese J Appl Meteorol Sci 21(3):287–297

Chapter 14
Lightning Detection

Abstract Real-time locations of lightning are now available at various levels of performance for all regions of the world. The modern era of lightning detection began in the 1970s with the development of ground-based sensors in a network using electronics and communications technology available at the time. That advance led to covering the United States with the National Lightning Detection Network within a decade, and similar networks have been deployed widely on national and regional bases. Long-range networks are available everywhere but give somewhat less precise lightning locations. Using a network, rather than a single sensor, is highly preferable. When using network data for human safety, users must be aware that some delay is inevitable from the detection to the appearance of the results on a device screen due to processing time and transmission delays. Satellite-based lightning detection is becoming more commonly available, although the lightning position locations may not be accurate enough for human safety decisions.

14.1 Overview

The modern era of real-time lightning detection began with basic research in the middle 1970s being conducted by several faculty and staff at the University of Arizona in Tucson, Arizona, in the United States. A primary motivation was forecasting the lightning danger at the Kennedy Space Center in Florida. They developed the technology for operating a network of ground-based sensors using the latest advances in electronics and communications that were available at the time (Krider et al. 1976). This approach was applied to forest fire detection soon afterward (Noggle et al. 1976; Krider et al. 1980). A huge range of demands for real-time and archived lightning data has taken advantage of this capability, including power utilities (McGraw 1982).

An average of 50 formal articles using data from ground-based lightning detection networks and 100 to 200 conference and other informal papers have been published per year in the last decade. A sampling of these publications in human safety, meteorology, and other topics are included in this book. The reference lists in those papers can be used to work back in time for the background leading to the present state of global recognition and implementation of lightning detection data.

M. A. Cooper, R. L. Holle, *Reducing Lightning Injuries Worldwide*,
Springer Natural Hazards, https://doi.org/10.1007/978-3-319-77563-0_14

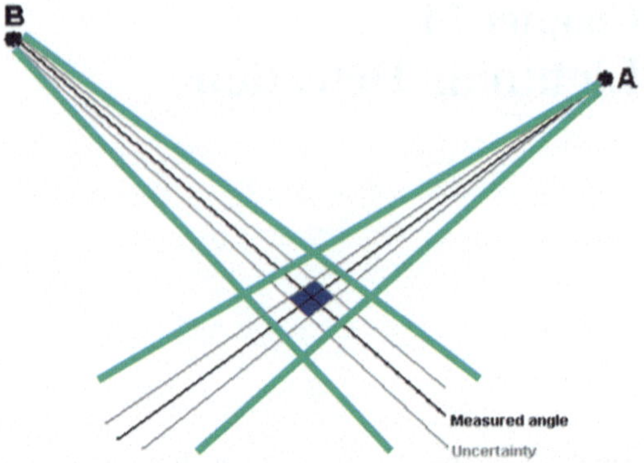

Fig. 14.1 Geo-location of lightning using direction finding. (Cummins and Murphy 2009; Rakov 2016)

14.2 Lightning Location Principles

When cloud-to-ground lightning contacts the ground, it emits an electromagnetic signal that is unique compared with all other man-made and natural signals. The vertical channel of cloud-to-ground lightning near the ground emits strongly in the low frequency (LF) to very low frequency (VLF) range. Lightning signals propagating in the LF/VLF range are termed as behaving as a ground wave in the earth-ionosphere waveguide and are capable of being used to locate lightning far beyond the line of sight (Nag et al. 2015). The cloud-to-ground electromagnetic signal can be defined on the millisecond time scale by a number of parameters that identify it as a lightning flash coming to ground (Krider et al. 1980; Rakov 2013). The initial detection technology measured the angle from the sensor to the ground strike location with a direction finder (DF). Angles from several sensors are sent to a central processor where sophisticated spherical trigonometric and statistical methods determine the most likely location and provide an error for each detected location (Fig. 14.1). This method is sensitive to errors in angle and range. Note that as the lightning impacting the surface of the earth is located farther away from the sensor, any error in angle contributes a growing location error.

After the DF approach was implemented, it was apparent that at long distances to a flash, angles alone had accuracy limitations due to random errors from various sources. The next step was to use the Global Positioning System (GPS) to provide very accurate timing to determine the time of arrival (TOA) when the electromagnetic signal from a cloud-to-ground event arrives at a sensor. These times determine the instant when the flash contacts the ground. The times are then calculated backward from the estimated ground strike locations to the sensors, and a statistical optimization technique is used to locate the flash. When the geometry of a network of three sensors is considered, however, it is found that two solutions are possible as shown by the shaded areas in Fig. 14.2. To avoid this ambiguous outcome, at least four sensors are needed for each light-

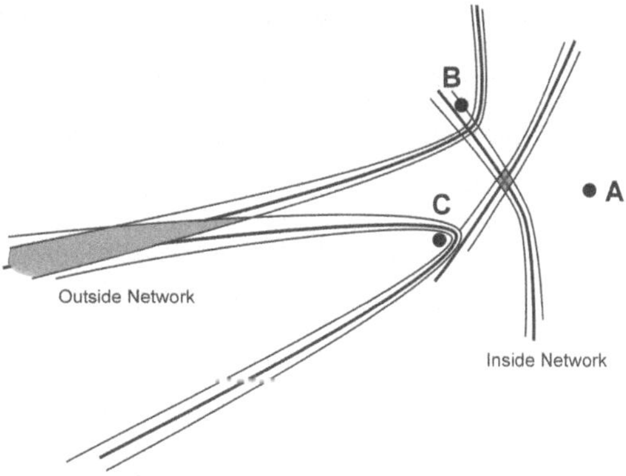

Fig. 14.2 Geo-location of lightning using time of arrival. (Cummins and Murphy 2009; Rakov 2016)

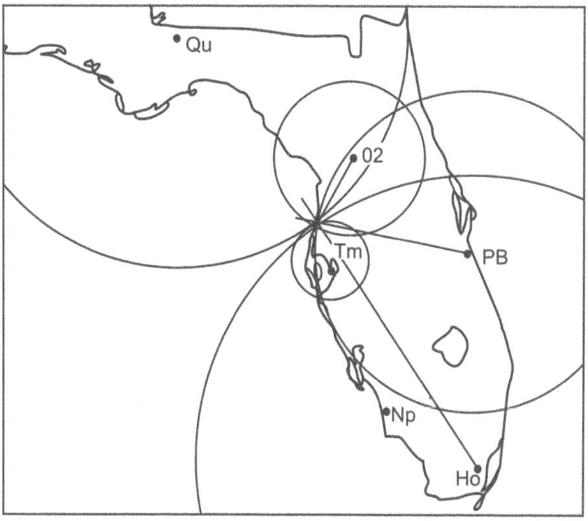

Fig. 14.3 Geo-location of lightning using both direction finding and time of arrival. (Cummins and Murphy 2009; Rakov 2016)

ning detection to have a unique location. Several types of networks using TOA only are described later (Fig. 14.3).

Question to Explore
What is the reason for the result of obtaining two locations with a three-sensor TOA network? Note: This is a rather complex geometrical concept involving hyperbolae.

Because of the limitations of DF-only and TOA-only networks, the combination of both methods was implemented by Vaisala in the mid-1990s (Cummins et al. 1998). The NLDN and GLD360 networks use both directions and times to locate lightning, as described in the following sections.

A ground-based network detects lightning with two or more sensors. The more times and/or angles that are available, the more accurate is the flash location. The time of the lightning event contacting the surface of the earth is accurate to a millisecond or better, but delays in disseminating the located solutions based on signals from multiple sensors are about 15 s in a national network and up to a minute or more in a global network. These delays are due to communications, since the central processor must wait to receive all relevant sensor reports in order to determine locations. Reliable real-time communications are typically the largest expense in establishing networks.

For common weather applications and warning systems that use data from such a network, other delays may be built into the system depending on the number of relays, sampling times, and other issues that affect the timeliness of the delivered data. While this time interval is not necessarily important for most public weather applications, delays need to be recognized by those using the results for safety, and they must be aware of what delays exist in their system.

In-cloud lightning can also be detected by LF/VLF networks, but in-cloud pulses emit more strongly in the very high frequency (VHF) range than in LF/VLF. As a result, VHF-sensor networks on a local scale have been developed such as the Lightning Mapping Array (LMA) in Sect. 14.7. However, lightning emissions in the VHF are only detectable within the line of sight, so that sensors need to be much closer together than in an LF/VLF lightning detection network (Sect. 14.7).

A complete summary of all lightning detection methods in recent years is provided by Nag et al. (2015). This comprehensive review includes the fundamental characteristics of lightning emissions to be detected, advantages and limitations at various frequencies, current ground- and satellite-based methods, detection efficiency and location accuracy calibrations as well as intercomparisons where available, and the types of lightning that each sensor detects.

14.3 Measurement Accuracy

Several measures of the quality of detection are routinely monitored. The location accuracy (LA) for each flash and stroke measures the distance between the network's position estimate and the actual position; zero kilometers is the ideal LA value. LA can be validated against ground truth, which is the actual known location where lightning occurs. Ground truth can be established using rocket-triggered lightning (Nag et al. 2011; Yijun et al. 2016), tall tower strikes (Warner et al. 2012), building strikes (Saba et al. 2017), and camera views (Stall et al. 2009).

Detection efficiency (DE) is the number of detected flashes or strokes divided by the actual number determined from ground truth; 100% is the desired DE. Signal strength, polarity, and the ability to classify correctly cloud-to-ground from in-cloud lightning are also measures of the quality of the data from a network and can be calibrated with rocket, tower strikes, and cameras.

14.4 National Lightning Detection Network (NLDN)

The NLDN began to cover the US mainland in 1989 (Cummins and Murphy 2009). This network consists of 119 sensors spaced across the 48 contiguous states, for which data have been collected for over 30 years. Cloud-to-ground flashes, cloud-to-ground return strokes, and in-cloud pulses are detected by the NLDN. On average, each cloud-to-ground flash is located by eight sensors providing both directions and times. Note that only two sensors are needed to locate a lightning event since there are two times and two angles that give redundancy to allow calculation of an unambiguous location. Two-sensor locations occur when a flash or stroke is weak or near the edge of the network. Signal strength and polarity are also identified, as well as measures of the accuracy of the location for each individual flash and return stroke.

LA within the NLDN has steadily improved to the present value of 200 m (660 feet) or less (Nag et al. 2015). DE now exceeds 90% for cloud-to-ground flashes and 70% for cloud-to-ground strokes. The polarity is determined correctly for over 90% of cloud-to-ground events, and signal strength estimates are within 20% of ground truth (Mallick et al. 2012; Nag et al. 2014, 2015). Due to these well-characterized, high-performance characteristics, the NLDN can be used to calibrate other networks (Abarca et al. 2010).

The sensors in the NLDN are spaced several hundred km apart so that no ionospheric reflection is encountered. When the LF/VLF signal emitted by lightning reaches the ionosphere at a horizontal distance of several hundred km from a sensor, the lightning signal becomes inverted and more convoluted. For this reason, only propagation paths not reaching the ionosphere are employed in the NLDN.

Rocket-Triggered Lightning

A well-established method to calibrate lightning detection networks and to test materials for the effects of lightning strikes is utilized at the Lightning Observatory at Gainesville and is operated by the University of Florida (Rakov 2016). For several decades, small rockets have been launched into overhead thunderstorms when they are about to produce lightning, with a high success rate (Rakov et al. 2014). Lightning is then initiated in the cloud at temperatures colder than freezing and travels down a trailing wire (that is destroyed) to an instrumented tower on the ground. Similar rocket-triggered lightning facilities are located elsewhere around the world for the same purposes (Ushio et al. 1997; Sun et al. 2014). Unfortunately, the Florida facility was closed in 2017 for lack of funding.

14.5 Additional National and Regional LF/VLF Networks

The Canadian Lightning Detection Network (CLDN) covers all of Canada south of the arctic tree line. The CLDN is owned and operated by Environment and Climate Change Canada and provides data for meteorological services as well as forestry (Burrows and Kochtubajda 2010; Kochtubajda and Burrows 2010). Its 84 sensors are connected with the NLDN in a network called the North American Lightning Detection Network (NALDN) to provide seamless coverage over the United States and Canada by sharing the angles and times from lightning events with each other's networks (Orville et al. 2011). Processing is done in Tucson, Arizona, for all NLDN and CLDN sensors. Degradation, which would normally be expected along the border between the countries if each network operated separately, is avoided.

Regional and national networks such as the NALDN are operational in more than 45 countries. The EUCLID (European Cooperation for Lightning Detection) network in Europe combines data from sensors in 24 national networks (Schulz et al. 2015; Azadifar et al. 2016; Poelman et al. 2016). A processing center in Austria takes the detections from 149 sensors (as of December 2014). The DE and LA within the interior of the network are similar to those of the NALDN. Other regional and national networks employ TOA only, such as the Earth Network Total Lightning Network over the United States, but have not been as intensively documented as the NLDN (Heckman 2014; Mallick et al. 2015; Li et al. 2016) that uses more sensors to achieve unambiguous positions in some situations. Sensor spacing and sensitivity vary among networks that result in differing DE, LA, signal strength accuracy, and the proper classification of cloud-to-ground versus in-cloud detections.

In many countries, data from national or regional networks are not distributed to the public by the network owners and operators, such as power utilities, forestry, defense, or meteorological agencies. The extra burden imposed by demands from non-paying external users can become viewed as negative in many situations.

14.6 Use in Safety Networks

There are many variations on this model, and some networks, such as EUCLID, show 1 h of lightning at http://www.euclid.org/realtime.html with a 2 h delay to be certain that the data are not used for commercial or safety purposes. A few national networks have been available on websites for a number of years. For example, the CLDN is shown at 10-min intervals at http://weather.gc.ca/lightning/index_e.html. The Hong Kong Observatory shows local hourly lightning at http://www.weather.gov.hk/wxinfo/llis/gm_index.htm. NLDN data for the United States are shown every 20 min with a 20-min delay at http://thunderstorm.vaisala.com/explorer.html. Other online sites have varying delays and coverage limitations that make them very unlikely to be useful for human safety. Great caution and awareness need to be given to understanding all of the documented and undocumented characteristics of the networks providing data that are shown at no cost.

14.7 Long-Range Networks

To detect lightning at distances of more than a few hundred km at LF/VLF, the ionospheric reflection becomes a benefit to be exploited rather than avoided, as with the NLDN and other national and regional networks. One, two, or more ionospheric reflections are used to locate the signal from very distant flashes, but the signal's features at those distances become more diffuse such that differentiating between cloud-to-ground and cloud flashes becomes increasingly difficult.

Additional degradations in DE, LA, polarity, signal strength, and lightning type (cloud-to-ground versus in-cloud) classification necessarily occur over increasing distances from the lightning event to the sensors. Nevertheless, the range of these highly sensitive antennas allows detection across the world. The Global Lightning Dataset GLD360 network (Fig. 11.2) uses both time and angle (Said et al. 2010, 2013; Pohjola and Mäkelä 2013; Poelman et al. 2013; Mallick et al. 2014). Another long-range system is the Arrival Time Difference Network ATDnet that uses a form of TOA only (Gaffard et al. 2008; Anderson and Klugmann 2014). In addition, TOA is used by the World Wide Lightning Location Network WWLLN (Rodger et al. 2006; Abarca et al. 2010; Hutchins et al. 2012; Rudlosky and Shea 2013).

DE for the GLD360 ranges from 35% to 75% and LA is from 2 to 5 km (1.25–3 miles) over the globe (Mallick et al. 2014). ATDnet and WWLLN over much of the globe have a DE from 5% to 15% with locally higher values, while LA is not well established in terms of performance at this time. A large amount of detail about these and other networks are provided in tables shown in the comprehensive summary by Nag et al. (2015).

14.8 VHF Networks

VHF networks have a very accurate LA of tens of meters and a DE of nearly 100% for cloud flashes. The most common installation is the Lightning Mapping Array (LMA) developed at the New Mexico Institute of Mining and Technology in the United States (Thomas et al. 2004). However, the DE for cloud-to-ground lightning is not very large since the vertical channel of a cloud-to-ground stroke does not emit well in the VHF. As mentioned in Sect. 14.2, VHF lightning emissions are only detected within the line of sight, which indicates that such networks cover much smaller areas than an LF/VLF network, such as the NLDN.

As mentioned in Sect. 14.3, LMAs are able to detect the full extent of long in-cloud lightning events that other ground-based networks are not capable of identifying. To date, measurements by LMAs have shown the detailed structure of the longest in-cloud flash that stretched for 321 km over Oklahoma and another with a temporal duration of 7.74 s over France (Lang et al. 2017).

LMAs often have at least twelve sensors, and only timing is used for detection (Krehbiel et al. 2000). As a result, LMA networks are potentially expensive to operate and thus are usually only deployed around regions on the scale of rocket launch facilities and research projects where high-precision lightning information beyond the ground strike position is desired.

Combinations of lightning detection methods are desirable in some settings. For lightning research and lightning protection at the Kennedy Space Center's space launch facility, an LMA is utilized to show the full three-dimensional structure of flashes aloft, while the ground contact locations are from an LF/VLF network (Preston and Fuelberg 2015). Examples of meteorological and lightning physics studies using the complex three-dimensional views from LMAs are in Krehbiel et al. (2000), Edens et al. (2012), Weiss et al. (2012), and Soula et al. (2015).

14.9 Satellite Detection

Detection of optical emissions by lightning can be made from satellites. Somewhat surprisingly, the brightness of the lightning signal is easily detectable with the sensor. It can differentiate lightning from bright highly reflective cloud tops, including at the equator in the middle of the day. The most commonly used sensors were the Optical Transient Detector and Lightning Imaging Sensor (Mach et al. 2007). These and some other sensors were in orbit for a number of years, although the coverage was restricted in area and time (Koshak 2010). Typically, detection occurred a few times a day over a subpoint below the satellite (Nag et al. 2015). National meteorological services and other agencies within countries, and combinations of such interests, are sponsoring a new series of satellites to cover separately the Americas, Europe and Africa, and Asia in the coming years (Goodman et al. 2013). The latest satellite for the Americas was successfully launched in 2017. These satellites require extensive calibration, and the data will be well validated before it is released for use, although it is typically provided at no cost.

Optically determined locations are less accurate than those obtained from ground-based networks due to the limitations of the optical sensors and the long distance from the satellite to the earth, compared with ground-based sensors. In addition, for lightning detection satellites that orbit at the equator, detection is best overhead at that location, with performance degrading somewhat away from the equator. Location accuracies of about 8 km (5 miles) are provided by current technology, which is not adequate for human safety for thunderstorms that are often no more than 10–20 km (6–12 miles) in diameter. Flash DE is nearly 100%, although currently it is difficult to distinguish between cloud-to-ground and in-cloud lightning using the emitted light. The complexity of comparing data from satellite pixels with ground-based detection networks is considered by Nag et al. (2015).

In addition, optical methods do not provide signal strength and polarity. As a result, while it is expected that satellite-detected lightning will be applicable for many uses to provide a larger perspective at little or no cost to the public, ground-based precision networks will be necessary for applications such as human safety and other situations where high spatial resolution is essential, such as public utilities and other applications.

14.10 Stand-Alone and Handheld Sensors

In addition to the proliferation of ground-based networks providing high-quality lightning location, stand-alone sensors, such as handheld devices, are occasionally used. Instead of relying on multiple sensors reporting times and/or angles, one sensor takes the signal strength and uses a general assumption between signal strength and distance to determine how far away the lightning is located. The signal strength of a cloud-to-ground flash or stroke has a range of over two orders of magnitude, from as small as 2–3 kiloAmperes to occasionally over 300 kiloAmperes.

The method assumes that a strong signal received at a stand-alone sensor indicates a nearby lightning event, while a received weak signal means the event is farther away. This is an important limitation since a weak flash close to the stand-alone sensor can be detected as the same intensity as a strong flash far away. The problem is somewhat mitigated by combining multiple strokes in an individual thunderstorm in order to have the collection of strokes indicate the distance. However, it is unreliable when the lightning event rate is low, and each intermittent flash needs to be used to make a distance estimate.

Some stand-alone sensors require a simultaneous light emission that is identifiable as coming only from lightning, but that can be limited by the variable range of detecting visible lightning. A stand-alone sensor usually reports directions in octants rather than a precise angle. Since exact angles and ranges are not detected by handheld devices, it is very difficult to develop reliable intercomparisons with calibrated ground-based detection networks. Since handheld devices are prone to the range conversion based on limited information available at one location, stand-alone sensors need to be used with great care for human safety (DeCaria et al. 2011).

References

Abarca SF, Corbosiero KL, Galarneau TJ Jr (2010) An evaluation of the worldwide lightning location network (WWLLN) using the National Lightning Detection Network (NLDN) as ground truth. J Geophys Res 115:D18206. https://doi.org/10.1029/2009JD013411

Anderson G, Klugmann D (2014) A European lightning density analysis using 5 years of ATDnet data. Nat Hazards Earth Syst Sci 14:815–829

Azadifar M, Rachidi F, Rubinstein M et al (2016) Evaluation of the performance characteristics of the European lightning detection network EUCLID in the alps region for upward negative flashes using direct measurements at the instrumented Säntis tower. J Geophys Res Atmos 121:595–606. https://doi.org/10.1002/2015JD024259

Burrows W, Kochtubajda B (2010) A decade of cloud-to-ground lightning in Canada: 1999-2008. Part 1: flash density and occurrence. Atmos-Ocean 48:177–194

Cummins KL, Murphy MJ (2009) An overview of lightning locating systems: history, techniques, and data uses, with an in-depth look at the U.S. NLDN. IEEE Trans Electromagn Compat 51:499–518

Cummins KL, Murphy MJ, Bardo EA et al (1998) A combined TOA/MDF technology upgrade of the U.S. national lightning detection network. J Geophys Res 103:9035–9044

DeCaria AJ, Wimer JW, Fijalkowski HM et al (2011) Detection efficiencies and range accuracies of three portable lightning detectors compared with the National Lightning Detection Network. In: Preprints of the 5th conference on the meteorological applications of lightning data. American Meteorological Society, Seattle/Washington. 22–27 January 2011

Edens H, Eack KB, Eastvedt EM et al (2012) VHF lightning mapping observations of a triggered lightning flash. Geophys Res Lett 39(19):L19807. https://doi.org/10.1029/2012GL053666

Gaffard C, Nash J, Atkinson N et al (2008) Observing lightning around the globe from the surface. In: Preprints of the international lightning detection conference. Vaisala, Tucson. 21-23 April 2008

Goodman SJ, Blakeslee RJ, Koshak WJ et al (2013) The GOES-R geostationary lightning mapper (GLM). Atmos Res 125-126:34–49

Heckman S (2014) ENTLN status update. In: Preprints of the XV international conference on atmospheric electricity, Norman. 15–20 June 2014

Hutchins ML, Holzworth RH, Brundell J et al (2012) Relative detection efficiency of the world wide lightning location network. Radio Sci 47(6). https://doi.org/10.1029/2012RS005049

Kochtubajda B, Burrows WR (2010) A decade of cloud-to-ground lightning in Canada: 1999-2008. Part 2: polarity, multiplicity and first-stroke peak current. Atmos-Ocean 48:195209

Koshak W (2010) Optical characteristics of OTD flashes and the implications for flash-type discrimination. J Atmos Ocean Technol 27:1822–1838

Krehbiel PR, Thomas RJ, Rison W et al (2000) GPS-based mapping system reveals lightning inside storms. EOS, 81:21–25

Krider EP, Noggle RC, Uman MA (1976) A gated, wideband magnetic direction finder for lightning return strokes. J Appl Meteorol 15:301–306

Krider EP, Noggle RC, Pifer AE et al (1980) Lightning direction-finding systems for forest fire detection. Bull Am Meteorol Soc 61:980–986

Lang TJ, Pédeboy S, Rison W et al (2017) WMO world record lightning extremes: longest reported flash distance and longest reported flash duration. Bull Am Meteorol Soc 98:1153–1168

Li D, Rachidi F, Rubinstein M et al (2016) Location accuracy evaluation of ToA-based lightning location systems over mountainous terrain. In: Preprints of the 24th international lightning detection conference. Vaisala, San Diego. 18–21 April 2016

Mach D, Christian H, Blakeslee R et al (2007) Performance assessment of the optical transient detector and lightning imaging sensor. J Geophys Res 112:D09210. https://doi.org/10.1029/2006JD007787

Mallick S, Nag A, Hill JD et al (2012) The NLDN performance characteristics: an update. In: Preprints of the 22nd international lightning detection conference. Vaisala, Broomfield. 2–3 April 2012

Mallick S, Rakov VA, Ngin T et al (2014) Evaluation of the GLD360 performance characteristics using rocket-and-wire triggered lightning data. Geophys Res Lett. https://doi.org/10.1002/2014GL059920

Mallick S, Rakov VA, Hill JD et al (2015) Performance characteristics of the ENTLN evaluated using rocket-triggered lightning data. Electric Power Syst Res 118:15–28

McGraw M (1982) On-line' lightning maps lead crews to 'trouble. Electr World 196:111–114

Nag A, Mallick S, Rakov VA et al (2011) Evaluation of U.S. National Lightning Detection Network performance characteristics using rocket-triggered lightning data acquired in 2004–2009. J Geophys Res 116:D02123. https://doi.org/10.1029/2010JD014929

Nag A, Murphy MJ, Cummins KL et al (2014) Recent evolution of the U.S. National Lightning Detection Network. In: Preprints of the 23rd international lightning detection conference. Vaisala, Tucson. 18–19 March 2014

Nag A, Murphy MJ, Schulz W et al (2015) Lightning locating systems: insights on characteristics and validation techniques. Earth Space Sci 2. https://doi.org/10.1002/2014EA000051

Noggle RC, Krider EP, Vance DL et al (1976) A lightning direction-finding system for forest fire detection. In: Proceedings of the 4th conference on fire and forest meteorology. American Meteorological Society, St. Louis. 16–18 November 1976

Orville RE, Huffines GR, Burrows WR et al (2011) The north American lightning detection network (NALDN) – analysis of flash data: 2001-2009. Mon Weather Rev 139:1305–1322
Poelman DR, Schulz W, Vergeiner C (2013) Performance characteristics of distinct lightning detection networks covering Belgium. J Atmos Ocean Technol 30:942–951
Poelman DR, Schulz W, Diendorfer G et al (2016) The European lightning location system EUCLID – part 2: observations. Nat Hazards Earth Syst Sci 16:607–616
Pohjola H, Mäkelä A (2013) The comparison of GLD360 and EUCLID lightning location systems in Europe. Atmos Res 123:117–128
Preston AD, Fuelberg HE (2015) Improving lightning cessation guidance using polarimetric radar data. Weather Forecast 30:308–328
Rakov VA (2013) Electromagnetic methods of lightning detection. Surv Geophys 34:731–753
Rakov VA (2016) Fundamentals of lightning. Cambridge University Press
Rakov VA, Mallick S, Nag A et al (2014) Lightning Observatory in Gainesville (LOG), Florida: a review of recent results. Electric Power Syst Res. https://doi.org/10.1016/j.epsr.2014.02.037
Rodger CJ, Werner S, Brundell JB et al (2006) Detection efficiency of the VLF world-wide lightning location network (WWLLN): initial case study. Ann Geophys 24(12):3197–3214
Rudlosky SD, Shea DT (2013) Evaluating ENTLN performance relative to TRMM/LIS. Geophys Res Lett 40:1–5
Saba MMF, Paiva AR, Schumann C et al (2017) Lightning attachment process to common buildings. Geophys Res Lett 44. https://doi.org/10.1002/2017/GLO72796
Said R, Inan U, Cummins KL (2010) Long-range lightning geolocation using a VLF radio atmospheric waveform bank. J Geophys Res 115:D23108. https://doi.org/10.1029/2010JD013863
Said R, Cohen MB, Inan US (2013) Highly intense lightning over the oceans: estimated peak currents from global GLD360 observations. J Geophys Res Atmos 118. https://doi.org/10.1002/jgrd.50508
Schulz W, Diendorfer G, Pedeboy S (2015) The European lightning location system EUCLID – part 1: performance analysis and validation. Nat Hazards Earth Syst Sci 16:595–605
Soula S, Defer E, Füllekrug M et al (2015) Time and space correlation between sprites and their parent lightning flashes for a thunderstorm observed during the HyMeX campaign. J Geophys Res Atmos 120. https://doi.org/10.1002/2015JD023894
Stall CA, Cummins KL, Krider EP (2009) Detecting multiple ground contacts in cloud-to-ground lightning flashes. J Atmos Ocean Technol 26:2392–2402
Sun Z, Qie X, Jiang R (2014) Characteristics of a rocket-triggered lightning flash with large stroke number and the associated leader propagation. J Geophys Res Atmos 119(23):13388–13399
Thomas RJ, Krehbiel P, Rison WS (2004) Accuracy of the lightning mapping Array. J Geophys Res 109:D14207. https://doi.org/10.1029/2004JD004549
Ushio T, Kawasaki Z, Ohta Y et al (1997) Broad band interferometric measurement of rocket triggered lightning in Japan. Geophys Res Lett 24(22):2769–2772
Warner TA, Cummins KL, Orville RE (2012) Upward lightning observations from towers in Rapid City, South Dakota and comparison with National Lightning Detection Network data, 2004–2010. J Geophys Res 117(D19). https://doi.org/10.1029/2012JD018346
Weiss SA, MacGorman DR, Calhoun KM (2012) Lightning in the anvils of supercell thunderstorms. Mon Weather Rev 140:2064–2079
Yijun Z, Lu W, Chen S (2016) A review of advances in lightning observations during the past decade in Guangdong. J Meteorol Res 30(5):800–819

Part IV
How to Reduce Global Lightning Casualties

Chapter 15
How to Make Baseline Studies of Lightning Deaths and Damages

Abstract Fatalities from lightning are more likely to be reported than injuries. National data are essential to ascertaining the level of lightning vulnerability of a country's population, and multiple years of data are highly preferable. Interannual fluctuations in lightning fatalities are to be expected due to the natural variability of lightning occurrence, and great care must be taken to avoid over-interpreting these changes. A governmental agency, such as a meteorological service, medical reporting system, or natural hazards reports should be considered as a first source of data. Reports need to include as much detail as possible. The ratio of injuries per death can be an indicator of the quality of a lightning casualty database.

With respect to damages, it is not possible to collect data about all of the direct and indirect impacts of lightning. The costs of long-term preventive measures to reduce the effects of lightning before it occurs may not show up in calculations. Industries usually do not release information on structural damages to the public, although anecdotes of spectacular stories may be useful in maintaining general lightning awareness.

15.1 Deaths and Injuries

Knowing the number of people killed and injured by lightning is critical to safety programs and awareness activities described in this book and can serve as a measure of the effectiveness of lightning safety education programs. Unfortunately, it may be difficult to collect a reputable database for fatalities and even more difficult for injuries in most countries.

First, what should be included? The data need to be collected uniformly over an entire country and over a period of as many years as possible. With this information, better strategies can be developed to reduce lightning casualties in a nation. Each case should include as much of the following information as possible in order to understand better the scenario of the people involved:

- Age and sex of each person
- Number of people killed and/or injured per event

M. A. Cooper, R. L. Holle, *Reducing Lightning Injuries Worldwide*,
Springer Natural Hazards, https://doi.org/10.1007/978-3-319-77563-0_15

- Village, city, district, and/or state
- Time of year
- Local time of day
- Weather at the time
- Location of the event relative to immediate surroundings
- Activity of the person relative to immediate surroundings

A government agency with uniform and consistent data collection methods is the most likely source of national-scale information. It is not apparent how any other type of organization can make a uniform cohesive data collection across an entire country. In the United States, the publication *Storm Data* collects information from each National Weather Service Forecast Office on all types of weather impacts, including lightning; such data have been collected by a standardized method for over 50 years (Curran et al. 2000). Despite this long period of record and an ongoing mandate to collect such data, as well as substantial media coverage of such events, some lightning fatalities, and especially injuries, can be missed due to relatively obscure media reports not reaching the meteorological forecast offices. An additional issue is that reporting systems may identify lightning as a secondary rather than primary cause of death or injury in the case of lightning starting a house fire, for example. In other countries, autopsy reports may be the only mandatory records of lightning fatalities that are collected throughout the country (Navarrete-Aldana et al. 2014). Elsewhere there may be datasets such as those related to disaster management (Illiyas et al. 2014; Badoux et al. 2016). All of these sources may have gaps in areal or temporal coverage or uneven quality that makes such data marginal or not acceptable.

15.2 Assessing the Quality of a Database

One should be careful to assess the many biases, both systematic and incidental, that can creep into a database. Major features of the dataset are described in the following subsections.

15.2.1 Ratio of Injuries per Death

One method to assess the quality of a database is to determine the ratio of injuries per death. An intensive US study in the state of Colorado found about ten injuries for every lightning fatality over a long period (Cherington et al. 1999). This 10:1 ratio is discussed in Sect. 6.3. It is also likely to apply to developed countries in Western Europe, Japan, Australia, and some other nations. A ratio smaller than 10:1 in developed nations can be considered indicative that not all injuries are being reported.

It is uncertain whether the 10:1 ratio applies in developing countries, since no rigorous injury data collection has taken place to date. In developing countries, there are frequent news reports of multiple deaths and injuries for each incident making it reasonable to consider the possibility that more fatalities and fewer injuries occur due to increased lightning exposure in schools that are not protected from lightning (Holle and Cooper 2016), during agricultural work (Holle 2016), and inside lightning-unsafe dwellings (Holle 2010). In these locations, multiple fatalities and injuries are more frequent than in the United States, so that the 10:1 ratio of injuries to deaths is not necessarily applicable.

There is another significant lack of information. Although the range of injuries is known clinically (Chap. 3), the distribution of survivor injuries across this spectrum is unknown and has not been collected on any extended population beyond the initial reports. When the ratio in developed nations reaches values lower than 10:1, it is possible that injuries are being missed, and perhaps the same is true in developing countries. It is also possible that some fatalities are also being missed. Although injuries are more difficult to determine, it is nevertheless important to collect injury data for these reasons:

1. The quality of the fatality data can be assessed with the injury-to-death ratio.
2. Data on injury age, sex, geographical location, activity, time of year and day, local weather situation, and location relative to surroundings are still very valuable in assessing the lightning danger and how to address education. This information is not typically dependent on whether a person is killed or injured.

15.2.2 Multiple Years Are Desirable

As many years of national data as possible are desired. If past national datasets can be found, they should all be analyzed, even if they are not up to date. Longer periods of data assist in developing more certainty about where, when, and under what conditions certain age and other categories of people are becoming casualties of lightning. Not all of the data may be adequate for every portion of the study, but they can be of some value in making some conclusions.

15.2.3 Geographical Coverage

Geographical coverage within a country needs to be uniform. It is possible that certain critical districts of a country will have missing reports for some of the years, particularly those areas that are difficult to reach or have poor communication systems. If this lack is not noted in the report, this can make the entire study unreliable. Particularly if one goal is to develop a national trend, one cannot make assumptions about what is occurring in parts of a country without great care, knowledge of the available database, and understanding of the situation.

15.2.4 Bias Toward Multiple-Casualty Events

Datasets are often biased toward multiple-casualty events because media and other reports tend to emphasize them. It is likely that many more single-fatality events are taking place that almost no one except the immediate family or co-workers knows to have occurred. A researcher should be careful not to overstate these multiple-casualty events as the normal or the common; they almost certainly are not.

15.2.5 Interannual Trends

One should be extremely cautious about trends from year to year. When a data collection method is new or has changed in any way from 1 year to the next, one should not speculate about the cause but should attempt to understand exactly how the data collection is being conducted, instead of making speculative explanations. For example, a study was recently published that rightly pointed out that in the last few years, many more fatalities were found than in previous decades, but the change was found to be almost entirely due to a much improved data collection method (Tilev-Tanriover et al. 2015). A recent study in Bangladesh showed that the expansion of mobile phone communications had greatly increased in the few years such that media were now aware of many more reports than had previously been the case (Dewan et al. 2017). In developing countries, higher fatality totals are being reported within the last 5–10 years that are almost entirely attributable to major improvements in reporting but are not actual trends in lightning casualties.

15.2.6 High-Profile Events

At times, specific high-profile events have occurred such as recent multiple-fatality totals that were reported over the span of a few days in Bangladesh and India (Holle and Islam 2016; Nag et al. 2017). These events have made the media, public, and data gathering agencies much more likely to report individual cases of lightning fatalities than had been the case in the past. Great care is emphasized in comparing fatality totals before and after such highly visible events that significantly change the data collection. Incidents with one or a few fatalities had been occurring all along but may be over emphasized when they are bundled together and capture the attention of the media; unfortunately, they do not represent a trend.

Since multiple-casualty events tend to be more frequently reported, it is also recommended to tally the number of lightning fatality incidents, regardless of the number of people killed or injured. The number of events should be somewhat more stable from year to year than the number of fatalities and injuries, so interannual changes in the number of incidents can assist in assessing the quality of the dataset.

An event count reduces apparent spikes in the number of deaths and injuries that occur during a multiple-casualty event that does not necessarily represent any important change from year to year. In developed countries such as the United States, where 90% of the events are to one person at a time, the fatality count is not much larger than the incident count. However, in developing nations, many events involve more than one person.

For example, many countries are dominated by mesoscale convective systems, called MCSs (Dotzek et al. 2005). These well-known meteorological features described in Sect. 13.2 cover very large areas. They have been measured to produce tens of thousands of cloud to ground flashes in a single night. Specific conditions are needed to produce MCSs, and the number of MCSs in a particular location varies naturally from year to year. A region or country may have only a few, or tens, of MCSs per year. The excess or deficit of MCSs in a year will affect the lightning occurrence for that year in a country, such that fatalities may be expected to fluctuate similarly as the flash count. As a result, a short-term increase or decrease in the annual fatality and injury count for a smaller country cannot be attributed to education or social-economic factors alone as the direct cause but to normal fluctuations in lightning occurrence. Many other meteorological phenomena naturally fluctuate from year to year, such as pre-monsoon and monsoon storms in Southeast Asia and the Indian subcontinent (Sect. 13.2).

Questions to Explore
What are some reasons why a one-year sample of lightning fatality data is not adequate? What about data on injuries? What about a partial year?

15.3 Damages

Chapter 5 described known damages. Finding the actual costs of lightning damage is nearly impossible! Lightning causes damage across a very wide spectrum of society and disciplines. There are direct causes, and those that are avoidance and mitigation costs. None of these is well identified.

The following are some examples:

- What is the cost of avoiding damage to a power line supplying power to a critical facility such as a hospital or a defense facility? It is not only the cost of the surge protection installed previously on the buildings but also that of the preplanned hardening of the incoming power to the buildings, the neighborhood, and the entire community, in addition to emergency generators made ready for power outages. While there may be no actual expenses at the time of the lightning event, the costs leading up to the event may be substantial and are usually undocumented separately by the industry that installed them.
- What is the total cost of a single passenger plane being delayed at an airport due to ground workers staying safe inside an airport building because lightning is a

threat? It is more than the minutes of time spent by the crew waiting for lightning to end. It involves delay to passengers on the plane, repositioning of aircraft on the taxiway, delayed takeoff of the aircraft that may miss its next flight later in the day, and lack of connections by pilots and other crewmembers who are delayed for their next flight. These costs can accumulate across the entire country or region and are very difficult costs to evaluate in actual expenses (Steiner et al. 2013).

No industry has developed a comprehensive evaluation of the actual costs of lightning, such as the two examples above. An important difficulty is that many impacts are self-reported and resolved; they can also be proprietary. In the case of a homeowner losing an electronic appliance during a thunderstorm, the insurance company is called and payment is made, but there is no central public clearinghouse for such events. Insurance companies are very reluctant to distribute such data outside their company due to concerns of revealing their cost and profit scales (Holle et al. 1996).

What can be done? There is a practical reason to attempt to collect such data. Direct causes of building fires, explosions, and other impacts can be collected on an anecdotal basis (Chap. 5). Political entities may appreciate that lightning is indeed a continuing source of expense and such large numbers may help attract their attention. However, actual totals are almost always are very much larger than documented costs.

There are only a few areas where data can be reasonably collected, and these are related to direct costs. Forest fires caused by lightning are well identified, and there often are exact figures for the direct expenses due to firefighting efforts (Chap. 5). Insurance companies may provide bulk data on lightning damage for an area for a year, but this is unusual. Utilities such as power companies and communications facilities may keep an internal record of damage due to lightning, although these usually will not account for the large avoidance costs of installing lightning protection on lines and substations, having repair crews on standby or maintaining an inventory of parts.

Attempting to collect data on lightning damage from outside a specific industry is not possible without their assistance. The best approach is to contact key facilities or affected companies and industries to determine if their data can be made available on a regional or national basis. These facilities will need to be convinced that revealing these costs to the public is also in their interest, but that is unlikely. Again, anecdotes are useful in gaining attention but never tell the full impact of lightning.

One can look at the United States *Storm Data* publication from the National Weather Service for an example of the difficulty. *Storm Data* is one of the premier databases on all types of weather impacts in the world and has been maintained for over 50 years. It is designed to tally the impacts of tornadoes, hurricanes, thunderstorms, floods, cold, heat, and all other weather-related hazards. However, in a study of entries from local NWS offices compared with actual insurance claims, a ratio of 367 actual insurance claims was found for every entry that made it into *Storm Data* (Holle et al. 1996) The NWS staff has no way of knowing about a single event damaging one homeowner's property, much less being able to identify the cost, time of day, and type of damage. An additional issue is that *Storm Data* does not include indirect impacts such as lightning-caused house fires and vehicle crashes due to winter weather. In summary, specific cases of damages from lightning can be pointed out as examples of the economic impacts of lightning, but the actual costs are unlikely to be attainable by an individual.

Can one spectacular lightning damage event such as an oil refinery explosion bring constructive attention to the frequent lesser damages of lightning? Can a public utility be convinced of the value of releasing their costs of lightning damage on an ongoing basis? Do such efforts result in appreciation for the need for lightning protection on schools and homes?

15.4 Maintaining Progress Statistics over Long Periods

After a database of fatalities and perhaps injuries has been developed, a system should be instituted to gather all of the statistics that are possible over subsequent years. Monitoring the data is necessary to make sure that the collection methods or quality has not deteriorated with time. In addition, homogeneous, standardized national data collection over many years can place the expected and normal year-to-year variations in proper perspective. It will also be easier to identify the at-risk population groups, how to address them, and how to take steps to reduce the toll of lightning on people.

At the end of each year, the findings should be evaluated and compared with the previous years. Are the numbers of people killed and injured by lightning consistent among the years? One should look at the number of events, age, number of people per event, gender, time of day and year, and similar basic statistics to be sure nothing has gone awry with the data collection. Data from one or a few years should never be used to make major shifts in safety campaigns or curricula.

Other pitfalls include:

1. For a smaller country, an annual national total in the single or low double digits will mean that a few incidents in 1 year will significantly affect the annual total – another reason to look at multiple years of data or pool data with other states, districts, or countries.
2. If a large number of fatalities occur in some events, it may appear that there has been a jump in the number of deaths in a particular year. A cross check would be to compare the number of fatality incidents from year to year to avoid the influence of unusual, multiple-casualty events.

Example

Consider a ten-year dataset for a country with the following sequence of fatalities per year: 47, 54, 44, 65, 48, 45, 58, 41, 56, and 42. The mean is 50 fatalities per year, and the range is from 41 to 65. Is the fact that the number of fatalities has decreased since the fourth year with 65 fatalities mean that all aspects of a lightning safety campaign are resulting in this difference?

References

Badoux A, Andres N, Techel F et al (2016) Natural hazard fatalities in Switzerland from 1946 to 2015. Nat Hazards Earth Syst Sci 16:2747–2768

Cherington M, Walker J, Boyson M et al (1999) Closing the gap on the actual numbers of lightning casualties and deaths. In: Preprints of the 11th conference on applied climatology. American Meteorological Society, Dallas. 10–15 January 1999

Curran EB, Holle RL, López RE (2000) Lightning casualties and damages in the United States from 1959 to 1994. J Clim 13:3448–3453

Dewan A, Hossain MF, Rahman MM et al (2017) Recent lightning-related fatalities and injuries in Bangladesh. Weather Clim Soc 9:575–589

Dotzek N, Rabin RM, Carey LD et al (2005) Lightning activity related to satellite and radar observations of a mesoscale convective system over Texas on 78 April 2002. Atmos Res 76:127–166

Holle RL (2010) Lightning-caused casualties in and near dwellings and other buildings. In: Preprints of the 3rd international lightning meteorology conference. Vaisala, Orlando. 21–22 April 2010

Holle RL (2016) Lightning-caused deaths and injuries related to agriculture. In: Preprints of the 33rd international lightning conference on lightning protection, Estoril. 25–30 September 2016

Holle RL, Cooper MA (2016) Lightning-caused deaths and injuries at schools. In: Preprints of the 33rd international conference on lightning protection, Estoril. 25–30 September 2016

Holle RL, Islam S (2016) Summary of multiple-fatality lightning events during mid-may 2016 in Bangladesh. In: Preprints of the 33rd international conference on lightning protection, Estoril. 25–30 September 2016

Holle RL, López RE, Arnold LJ et al (1996) Insured lightning-caused property damage in three western states. J Appl Meteorol 35:1344–1351

Illiyas FT, Mohan K, Mani SK et al (2014) Lightning risk in India: challenges in disaster compensation. Economic & Political Weekly, 7 June, XLIX:23–27

Nag A, Holle RL, Murphy M (2017) Cloud-to-ground lightning over the Indian subcontinent. In: Postprints of the 8th conference on the meteorological applications of lightning data. American Meteorological Society, Seattle/Washington. 22–26 January 2017

Navarrete-Aldana N, Cooper MA, Holle RL (2014) Lightning fatalities in Colombia from 2000 to 2009. Nat Hazards 74:1349–1362

Steiner M, Deierling W, Bass R (2013) Balancing safety and efficiency of airport operations under lightning threats: a look inside the ramp closure decision-making process. J Air Traffic Control 55:16–23

Tilev-Tanriover S, Kahraman A, Kadioglu M et al (2015) Lightning fatalities and injuries in Turkey. Nat Hazards Earth Syst Sci 15:1881–1888

Chapter 16
Identification of Safe and Unsafe Areas

Abstract Three underlying principals need to be recognized with regard to lightning safety: (1) Assume that a location is unsafe until it can be shown to be made safe from lightning, (2) Being safe from rain is not the same as being safe from lightning, (3) Myths about lightning safety are more often wrong than correct.

16.1 Assume a Place Is Unsafe

There are only two reliable places to be safe from lightning. One is inside a large substantially constructed building; the other is inside a fully enclosed metal-topped vehicle. These locations provide an effect similar to a Faraday cage such that lightning striking a building or vehicle travels around, rather than through the people inside it.

Buildings that are safe from lightning have paths for lightning to follow through grounded wiring and plumbing in the walls that are properly installed according to accepted municipal building codes, and may have metal structural members that are part of the building itself (Holle 2010). Direct strikes to dwellings with people inside are quite common in developed countries but rarely result in fatalities, and injuries are mostly minor due to people being in contact with the conducting paths of wiring and plumbing (Chap. 8).

Any other building should be assumed to be unsafe - especially small structures. It is possible to make them lightning-safe, but it takes a specialized effort by a licensed experienced specialist in lightning protection and incurs expenses that may not be cost-effective or practical (Kithil and Rakov 2001). Any small and/or open-sided structure of any type should be assumed to be unsafe. These include agricultural outbuildings, shacks and huts, roadside shops, sun shelters, beach shelters, rain shelters, golf shelters, bus shelters, forest huts, and similar small enclosures.

Although many buildings in developing countries may appear to be "substantial," not all may have the protective metal cage of plumbing, wiring and metal building components coursing through the walls. An exception to this may be the small shops that are nearly ubiquitous along the main roads of developing countries. If they are housed in discarded metal shipping containers, there is a possibility that they may be at least somewhat protective from lightning injury.

M. A. Cooper, R. L. Holle, *Reducing Lightning Injuries Worldwide*,
Springer Natural Hazards, https://doi.org/10.1007/978-3-319-77563-0_16

Vehicles that are safe are fully enclosed and have metal tops. A direct strike can be very disconcerting, but one will survive a lightning strike while inside them (Holle 2008). In principle, they are portable lightning safety locations that can be moved to critical locations when the threat of a thunderstorm arrives (Chap. 7). Lightning-unsafe vehicles include golf carts, cloth-topped, four-wheeled, and similar open-sided vehicles.

Questions to Explore
How safe from lightning is a three-story hospital? A dune buggy? Can one tell from looking briefly at a building if it is lightning-safe?

16.2 Rain Avoidance Is Not Lightning Safety

Most lightning occurs in the presence of rain, but the first reaction is to pay attention to the rain rather than the lightning. Rain causes discomfort and makes people and objects wet but does not normally injure unless flooding occurs or the person becomes hypothermic. In contrast, lightning, while very intermittent in time and space, is potentially fatal.

Around 10% of all lightning casualties in developed countries occur under trees from sideflash or other mechanisms (Chap. 2). The immediate reaction is to try to stay dry under a tree, but simultaneously a very real lightning threat occurs here (Makela et al. 2003; Holle 2012). The percentage of people killed and injured under trees has been in the 10% range for decades, in all countries. Animals also tend to huddle under trees from the rain and thereby have a significant risk from lightning, as observed from the large number of reports of domestic and wild animals found dead under trees (Chap. 5).

Similarly, tents provide rain protection but are not lightning-safe. The problem is compounded when thunderstorms occur at night and people are sleeping flat on the ground, increasing their chances of injury both from relative tent height and isolation and ground current (Holle 2014; Chap. 2). Similar situations include going under open-sided stands at soccer or other recreation venues to escape rain while the lightning threat prevails (Holle 2005).

An additional danger is posed by open-sided structures in the middle of agricultural fields (Holle 2016). They may be provided for agricultural workers to stay dry during an afternoon storm. However, unless they have been specifically and correctly designed to protect from lightning by a licensed, experienced specialist in lightning protection, they should be assumed to provide no protection whatever from the lightning threat for those who were primarily avoiding rain.

16.3 Myths

Myths abound in all aspects of meteorology, and nearly all of them are incorrect (Chap. 9). Anecdotes are told and retold about storms missing here and there, never affecting certain places, or being associated with some irrelevant event. When an

Table 16.1 Common lightning myths

Wearing red attracts lightning
Mirrors should be covered when thunderstorms are in the area
Standing under power or telephone lines offers protection because the wires will preferentially be hit
Building your house next to certain trees or avoiding certain trees is protective
The rubber tires are what protects a person in a metal vehicle
"Dry lightning" is not dangerous
Lightning does not occur outside of the rain area
Metal attracts lightning
Crouching down lowers your chance of being injured because it lowers your height

event is intermittent or occurs sporadically at a specific place, there is a tendency to make up myths to fit that specific case, but they provide no overall solution and are not generalizable. These stories often become embellished or eventually morph into cultural "facts."

Lightning is no different. The suddenness and seemingly random nature of lightning leads to endless uninformed conjecture. Often it is thought, based on no factual or scientific information, that if a person doesn't do this or do that, they will be safe (Table 16.1).

Many myths are related to the mistaken assumption that lightning fatalities and injuries result from a direct strike coming downward from above a person. As outlined in Sect. 2.2, there are five mechanisms of lightning effects on people. One of the most commonly heard conjectures is that lowering one's height by going into a small ravine or crouching down. After burning its way through a kilometer or more of air, it is hardly reasonable to assume that lowering one's height by ½ m will make a difference. First of all, only about 3–5% of all lightning effects on people are due to the direct strike, where the height of the person is a consideration (Table 19.3). One study showed that squatting or lowering one's height was calculated to lower one's changes of injury by only 1.5–2.5% (Roeder 2014). Ground current, side flash, direct contact and upward leaders are all more frequent mechanisms of injury.

Other conjectures include thinking that standing on a rubber mat, wearing foam shoes, or standing a certain number of feet away from some object or that trees of certain types will make a person safe. These are potentially deadly guesses. This sort of approach is actually due to the fact that around 90% of all people affected by lightning survive, at least in the developed nations. The medical reasons are described in Chaps. 2 and 3. Nevertheless, the result of this high survival rate, often with devastating and permanent disabilities, is that stories are made up about exactly what a specific person was doing or wearing, or where they were standing, or where they were located relative to other objects for a particular instance. Explaining survival based on incidental irrelevant issues that were present during a single event is not a sound approach. Stories that are imagined to fit that one-time situation should never be the basis for safety advice. Unfortunately, those stories are repeated and enhanced until the conjectured reasons for surviving lightning become misguided, but repeated, recommendations.

Questions to Explore

What conjecture have you heard about how to avoid being killed or injured by lightning? Does it have any basis in fact? How can you tell which are true from those that are not?

References

Holle RL (2005) Lightning-caused recreation deaths and injuries. In: Preprints of the 14th symposium on education. American Meteorological Society, San Diego. 9–13 January 2005

Holle RL (2008) Lightning-caused deaths and injuries in the vicinity of vehicles. In: Preprints of the 3rd conference on meteorological applications of lightning data. American Meteorological Society, New Orleans. 20–24 January 2008

Holle RL (2010) Lightning-caused casualties in and near dwellings and other buildings. In: Preprints of the 3rd international lightning meteorology conference. Vaisala, Orlando. 21–22 April 2010

Holle RL (2012) Lightning-caused deaths and injuries in the vicinity of trees. In: Preprints of the 31st international conference on lightning protection. Vienna. 2–7 September 2012

Holle RL (2014) Diurnal variations of NLDN-reported cloud-to-ground lightning in the United States. Mon Weather Rev 142:1037–1052

Holle RL (2016) Lightning-caused deaths and injuries related to agriculture. In: Preprints of the 33rd international conference on lightning protection, Estoril. 25–30 September 2016

Kithil R, Rakov V (2001) Small shelters and safety from lightning. In: Proceedings of the international conference on lightning and static electricity. Society of Automotive Engineers, 2001-01-2896, Seattle/Washington. 10–14 September 2001

Makela J, Karvinen E, Porjo N et al (2003) Attachment of natural lightning flashes to trees: preliminary statistical characteristics. J Light Res 1:9–21

Roeder WP (2014) Backcountry lightning risk reduction-lightning crouch versus standing with feet together. In: Preprints of the 5th international lightning meteorology conference. Vaisala, Tucson. 20–21 March 2014

Chapter 17
Lightning Protection

Abstract The purpose of lightning protection systems is to protect structures and, to some extent, the people within them, as well as power systems, electronics, and other infrastructures from damage by lightning strike as well as by electromagnetic transients that may accompany lightning strikes. This chapter will discuss the basics of lightning protection systems and its history, proposed modifications that have not proven to live up to their claims, and pose ethical questions where providing partial protection may be the best decision for available resources. Surge protection will not be covered.

17.1 Introduction

As discussed in many of the previous chapters, lightning can cause significant damage not only to people and livestock but also to buildings, utility systems, industrial installations, banking, sensitive electronics, aviation, military and naval operations, shipping, forestry, and many more industries. Lightning causes fires, downtime, costly repairs, structural damage, power interruptions, and data losses, all of which can result in significant economic loss, particularly in developing countries.

"Lightning protection" (LP) is different from lightning safety or lightning injury prevention. Lightning protection nearly universally means protection of structures. While international codes take into account human injury, few lightning protection codes claim to protect people or animals within the buildings, especially small buildings. Obviously, if prevention were possible, most people, businesses, and governments would prefer to minimize lightning damage. Most developed countries have building codes that specify lightning protection for certain structures, usually depending on their function such as a school or hospital. In the developing world, even if a code is specified by the nation's bureau of standards, it may not implemented at the construction level because of lack of familiarity with the codes, lack of experienced protection engineers, availability of code-compliant materials, and especially the expense where income may be only US$1 to $10/day per person. For villagers who construct their own homes, lightning protection is seldom part of their knowledge base.

M. A. Cooper, R. L. Holle, *Reducing Lightning Injuries Worldwide*,
Springer Natural Hazards, https://doi.org/10.1007/978-3-319-77563-0_17

It is not the purpose of this chapter, nor this book, to discuss lightning protection system (LP) design. There are widely available extensive publications on the topics such as those by Golde (1973), FAA (1990), Rakov and Uman (2003), Uman (2008), and Cooray (2010), and there are internationally recognized lightning protection codes, most notably International Electrotechnical Commission (IEC 62305–1,2,3,4). Proper lightning protection design is complex and must be customized to the structure being protected depending on:

1. The function that the building serves such as a financial institution, school, munitions factory, hospital, or a church
2. Whether people or only structures are to be protected
3. The risk/benefit of protection versus replacement of an unprotected structure and other considerations

This chapter will cover not only what a lightning protection system consists of and its function but also some of the reasons that most people, businesses, and governments find judging claims of lightning protection systems to be confusing and arduous. There are well-recognized codes that should serve as the basis for the design of lightning protection whenever it is considered. Only trained specialists who design lightning protection systems as their daily work should be employed. Most electrical engineers, electricians, and others have little or no real knowledge of lightning protection and lightning protection codes regardless of the term "electric" appearing in their title. This chapter will also consider current controversies in lightning protection and "junk" science employed by some manufacturers.

17.2 Parts of a Lightning Protection System

There are three basic parts of any effective lightning protection system that are illustrated in Fig. 17.1:

1. *Air terminals*, commonly called lightning rods or arrestors, which intercept or serve as an "attachment" point for lightning, the first part of diverting the lightning from damaging the structure
2. *Down conductors* that connect the air terminals to a grounding system in order to harmlessly channel the lightning energy around a structure from the air terminals to the ground or, in the case of boats, water around a structure
3. *Ground terminals* or electrodes, also called earthing, which effectively dissipate the lightning energy into the ground or water and away from the structure and its contents

This three-part system serves to divert the energy and shield a structure from damage. To some extent, it will also minimize the electric and magnetic fields within the structure that are generated by lightning. If the structure to be protected contains electronics, communication and power systems, or other sensitive equipment

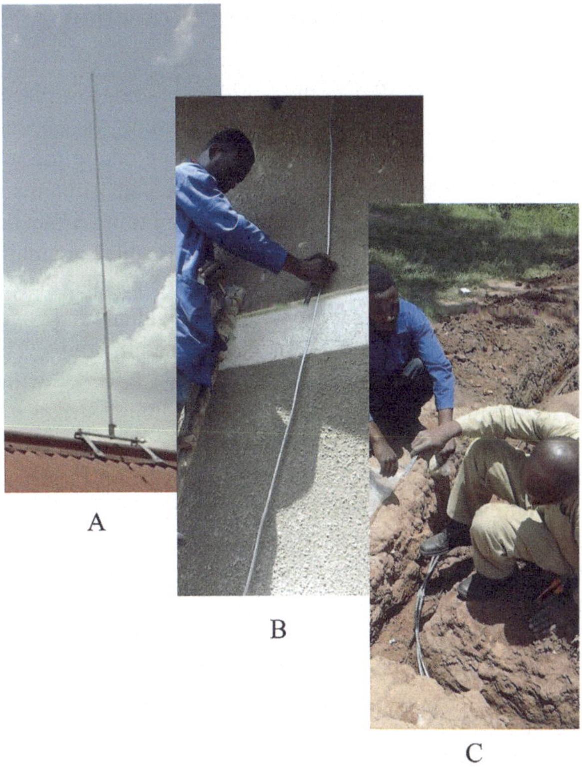

Fig. 17.1 Installation of (**a**) air terminal, (**b**) down conductor, and (**c**) ground terminal at school in Uganda. (Courtesy of ACLENet)

that need protection, surge protection may also be necessary. Surge protection will not be discussed in this book.

17.3 How Lightning Protection Works

Originally, people thought that lightning protection would *prevent* lightning strikes, including Benjamin Franklin, who invented the first effective lightning protection in 1752. Various theories were developed that the charge from the ground would somehow travel upward through the system and leak into the air above the structure, slowly dissipating atmospheric charges to prevent the rapid and often violent lightning strike.

Observers soon learned that lightning would strike "protected" buildings in spite of LP installation. As a result, they began studying the best ways to conduct the energy from the inevitable lightning strike harmlessly around a structure using the three-part system: air terminals, down conductors, and ground terminals.

17.4 Barriers to Lightning Protection

Lightning occurs all over the world, in high or low places and in nearly all climates, except for Antarctica (Chap. 11). Lightning has been called the weather hazard "most commonly experienced by most people in the world." Unlike other weather hazards such as hurricanes, tornadoes, and tsunamis, damage from lightning is largely a preventable risk with proper lightning protection.

However, until a disaster happens where there are multiple deaths, an extensive fire or a power outage affecting thousands of people, there is often a lack of foresight based on the rarity of the event. Both individuals and governments usually have issues that are of much more concern to them than lightning. There may be a general fear of electricity and lack of knowledge of how LP works, with some continuing to believe that lightning protection attracts more lightning strikes to a building or an area.

Some may feel that lightning injury is an "act of God," inevitable and unpreventable, or that it is sinful to attempt prevention. In many cultures, it is believed the lightning can be called down by witches to punish an enemy. Still other communities feel that a family who has suffered a lightning incident is cursed, and the family may be shunned, forcing them to move away and start over in a place where they are not known. Still others will believe, or at least act like, lightning disaster will never happen to them.

Lightning protection codes may be perceived as too complicated, and there is often a lack of qualified designers and installers. LP costs may be prohibitive compared to other priorities. Those who have done lightning protection with copper or aluminum may have had repeated thefts, with expensive replacement costs.

17.5 Risk Reduction and Lightning Protection Codes

Lightning protection is required by building codes in many parts of the world and almost unknown in others. The most widely accepted international code for lightning protection, IEC 62305–1,2,3,4 (International Electrotechnical Commission), lists four types of loss that are to be considered in determining the level of LP design:

- L1: loss of human life (including permanent injury)
- L2: loss of service to the public (such as utilities, power and communications, aviation)
- L3: loss of cultural heritage
- L4: loss of economic value (structure, content, and loss of activity)

Risk, depending on the lightning density and many other factors, can be assessed for each of these (Table 17.1). The need for lightning protection is often specified for different industries, in building codes, and other resources. Protection may be needed for industrial parks, manufacturing plants, churches, schools, banking centers, hospitals, military installations, historic landmarks, emergency centers, sport-

Table 17.1 Risk assessment factor assigned to the level of protection required for different situations (International Electrotechnical Commission)

Protection level	Current (kA)	Energy (kJ/Ω)	Efficiency (%)	Type of damage	Risk assessment factor
I	200 or more	10,000	98	Loss of lives (high human level concentration)	10^{-5}
II	150	5600	95	Loss of essential public services (telecommunications)	10^{-3}
III	100	2,500	90	Loss of cultural assets (monuments)	10^{-3}
IV	<100	<2500	80	Areas with low human presence, no public services, and no cultural interests	10^{-2}

Fig. 17.2 Typical rural housing in Africa and many other developing countries, consisting of mud brick walls with roofs of generations-old, tinder-dry thatch or sheet metal held down by stones. The walls do not contain wiring or plumbing, and these buildings are not considered lightning safe. (2015 Zambia photo ©M.A. Cooper)

ing complexes, correctional facilities, corporate centers, chemical plants, oil refineries, nuclear plants, and many other facilities.

Although petitions have been made recently to the IEC for development of more specialized LP codes, currently they do not exist for:

1. Small structures, including homes, commonly found in rural areas (Fig. 17.2)
2. Protection or people and animals, especially in fields and other open areas
3. Small boats, which may double as a family's home in addition to work area in some parts of the world

17.6 Ethical Versus Practical Considerations

In developed countries, it is easy to insist on nearly 100% assurance of safety by recommending substantial housing and fully enclosed metal vehicles as safer areas when thunderstorms are in the area and to vehemently dismiss partial measures.

However, in developing countries where families are often at risk 24/7/365, is it reasonable and ethical to compromise for partial protection that may save some lives? These topics are addressed by Rakov (2000), Hartono and Robiah (2007), Kumarasinghe (2008), Gomes (2010), Gomes and ab Kadir (2010), and Gomes et al. (2012). For example:

1. For developing countries, code-compliant LP designs may cost more than a family's yearly income. Can less expensive materials be used that will still provide adequate protection? Is there a significant loss in the safety margin?
2. Grounding or earthing of lightning can be incredibly difficult under the conditions of the dry seasons in some countries. Is it permissible to protect structures well for the wet seasons, when thunderstorms are more likely, but sacrifice the quality of the grounding systems in the dry season, when there is little chance of lightning or thunderstorms, because adequate year-round grounding in some areas is so difficult to achieve? Is partial safety better than none at all?
3. The integrity of the LP installation and ground resistivity should be assessed at least every 2 years. What are the ethics, versus the reality, of installing lightning protection systems but not being able to ensure regular maintenance and testing or of communities not being able to maintain lightning protection on their homes and schools due to its expense?
4. Can villagers be taught to install lightning protection, to use alternative materials to decrease cost, and to pass the instruction on to others accurately? Alternatively, is this a recipe for disaster if standards are not met, connections for the down conductors are not followed or understood by the villagers, or other materials or installation errors are made?
5. How many lives could be saved by promoting the use of rubber-soled shoes or sleeping mats and mattresses to at least partially protect from ground current in developing countries, even though they do not meet the standards that we insist on in the United States and other developed countries, and scientific testing has not been made as to their value?

Questions to Explore

Some testable projects are the following, courtesy of John Gookin, Ph.D., an experienced survivalist in such locations as Africa, Patagonia, and the Arctic: How much would one-half inch of closed cell foam insulation (flip-flops) below a standing human reduce ground current effect? How much does

(continued)

one-half inch of closed cell foam insulation (thin sleeping pad) below a sleeping human reduce ground current effect? What would be the effect of a circular conductor, such as a chain, loosely laid near the ground/soil surface and pinned to the ground with small metal tent stakes or bent nails, in reducing ground voltage differential to a human standing or lying down inside of it? Does a person sitting on bare ground with random hand and foot placement receive less ground current than a person lying flat on the ground with random foot and hand placement?

17.7 Difficulties Evaluating Product Claims: Frauds and Fakes

As in any area requiring highly specialized technical expertise, the public, governments, and industry may not have a knowledge sufficient to judge the adequacy of lightning protection design, installation, or choice of materials. Unfortunately, there is:

1. A lack of a perceived need for lightning protection resulting from a supposed rarity of strikes
2. The lack of engineers and architects trained in the very specialized nature of lightning protection engineering
3. Often the lack of industry standards governing LP purveyors and installers

As a result there is no "common knowledge" by the public or by most industry or governments that would allow them to temper the "scientific" claims, much less judge bids or discover the price gouging that can occur, including marketing of outright frauds and fakes (Uman and Rakov 2002).

Sometimes, the simplest is the best. The old-fashioned Franklin rod, plain, simple, and straight, with no arms, radioactivity, brushes, balls, colored globes, or other additions, remains the standard for lightning rods, with all others tested against it. However, there have been many challenges to this approach.

Table 17.2 lists some of the names used by purveyors of these scientifically unproven systems. Because many of these names are specifically condemned by internationally accepted lightning protection codes, sellers and manufacturers have attempted to deceive buyers by frequently changing the names and acronyms they use. Anyone seeking lightning protection should be wary and check out the claims of sellers.

17.7.1 Early Streamer Emitters: The Theory Versus the Reality

In the 1960s and 1970s, due to the high price of copper and other metals used in lightning protection, some proposed a technique that was said to reduce the number of air-terminations and down conductors that were prescribed in the international standards.

Table 17.2 Some names of unproven lightning protection systems

Early streamer emitters (ESE)
Lightning eliminators
Charge dissipation Array (CDA)
Dissipation Array systems (DAS)
Charge transfer system
Spline ball ionizer
Lightning suppressor

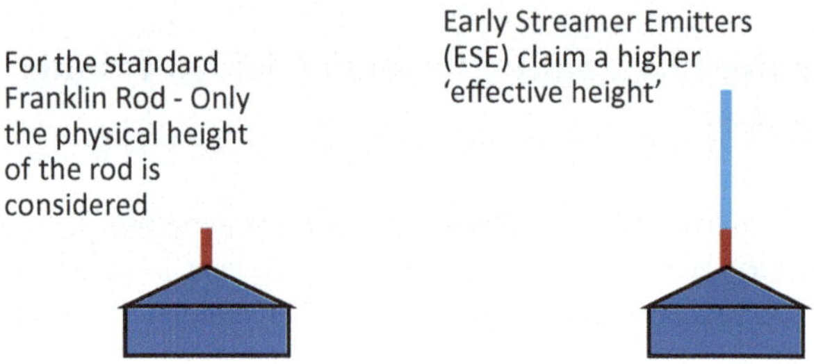

Fig. 17.3 Due to the addition of components that were claimed to emit upward streamers earlier to meet the down-coming lightning leader, it was hypothesized that ESE rods would have a higher "effective height" than the old-fashioned Franklin rod that was the recognized standard. (Courtesy of C. Gomes)

The Early Streamer Emitters (ESE) concept added components to the air terminal that were theorized to launch an upward streamer/leader much earlier than the conventional Franklin rod did. This terminal would attach to the downward leader earlier, making the "effective height" of the rod "taller" than it actually measured (Fig. 17.3). Being taller enabled a wider radius of coverage, necessitating fewer rods and down conductors to cover the area of concern, theoretically making the system less expensive (Fig. 17.4).

Variations of the ESE to induce the streamer included adding radioactive sources, electronic devices that were to inject voltage pulses, piezoelectric devices that were to generate voltage pulses, complex electrode systems, and others where the electric field was supposed to be modified by the shape of the tip. Figure 17.5 shows a variety of ESE variations. Other modifications included different shapes, adding arms or spikes, metal balls, golden metal coatings, colored glass globes, and all sorts of visual candy that added nothing to the function of the ESE.

The scientific community had doubts about these claims and would not accept them until the claims were tested. Independent testing did not support the ESE claims, and researchers stated that any slight effect of an ESE was not significant enough to justify the reduction in the number of rods.

The US lightning protection standards, codified and named after the National Fire Protection Association (NFPA780), accepted only the Franklin rod. A panel

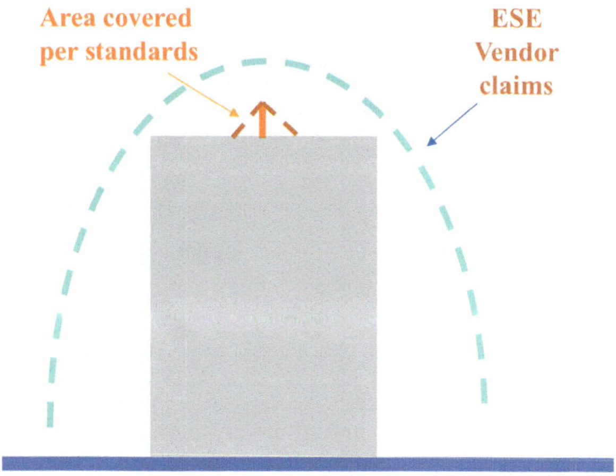

Fig. 17.4 ESE manufacturers claimed that by having a higher effective height, a larger radius of coverage was achieved, necessitating fewer rods to cover the area to be protected. This was proven to be a false claim in independent scientific trials. (Courtesy of C. Gomes)

was appointed in the early 1990s to investigate the possibility of including ESE in the standards (Draft NFPA781). In 1993, the panel declined to approve the proposed draft NFPA781, but ESE makers responded by taking the matter to court, claiming NFPA781 had just as much scientific support as the Franklin rod. In a settlement, the NFPA agreed to have ESE technology reevaluated by an outside panel. This independent panel confirmed that there was no scientific basis for NFPA 781, and ESE claims were again rejected. Eventually, ESE manufacturers and purveyors packed the NFPA committee and voted to include the unsupported claims so that NFPA has ceased to be a respected lightning code. Unfortunately, ESE devices were included in the French National Standards (NF C17–102) in the mid-1990s. Spain followed France and adopted ESE concept in their standards as well. As a result, the French and Spanish codes are NOT accepted standards.

While the ESE is no better than the plain Franklin rod of the same height, it is no worse either. The problem arises when the sales persons tell the buyer that they can save them money with their *improved* technology by decreasing the number of air terminals and down conductors. However, sellers often make up more than the difference by charging more for the fancy knobs, points, radioactivity, or shiny balls that are part of the claimed "improved" technology of the ESE design. Sophisticated marketers will use the same calculations and statistics as the standards but then substitute their shinier ESE and charge more than a standard Franklin rod installation. Others, who understand very little of the physics, will claim that ESE's avert or prevent lightning strikes by "discharging the electric field in the thundercloud." Extensive crucial evaluation of the ESE design is found in Becerra and Cooray (2007, 2008), and Hartono and Robiah (2006).

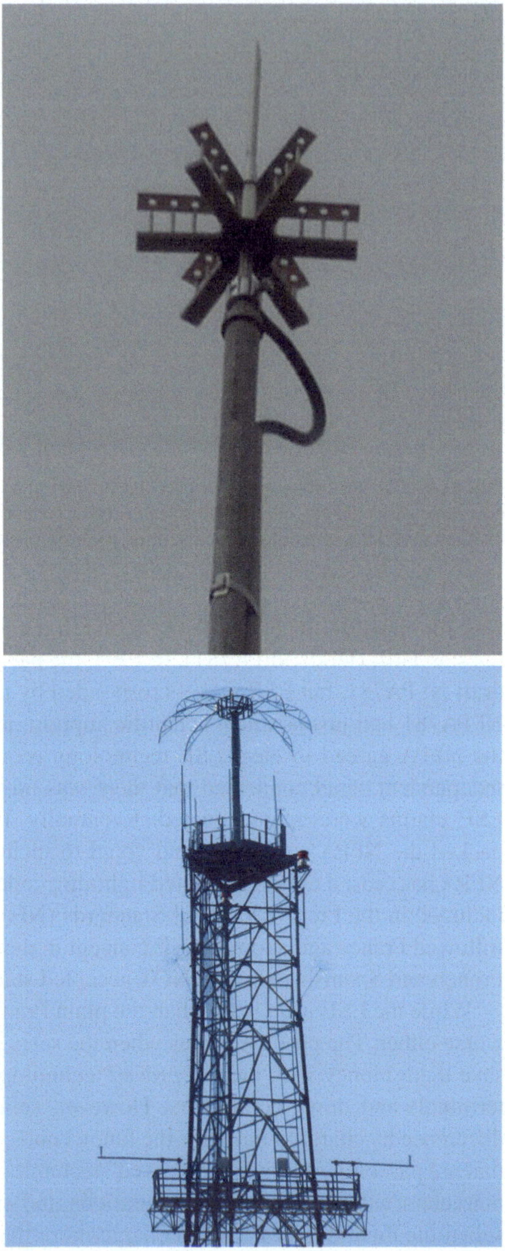

Fig. 17.5 Two ESE air terminal designs. (Photos courtesy of the National Lightning Safety Institute)

17.7.2 *Lightning Eliminators: Not Accepted by Any Code*

Another device proposes to repel or neutralize a lightning leader so that it causes no harm to the protected site. A "charge dissipation array" is used instead of a standard air terminal (see Fig. 17.2 for other names). Sellers of this product claim that a large amount of space charge is rapidly generated when a stepped leader is present, and either the opposite charge in the leader is neutralized or the "large space charge" repels the downward leader, preventing or eliminating the lightning strike. These "repellers," "dissipators," and "eliminators" are easily recognizable for their dozens or hundreds of fibers (Fig. 17.6). Not a single standard in the world has accepted this concept, but they are still frequently marketed to the unwary buyer.

17.8 Building Resilience/Decreasing Costs

It is unfortunate, but true, that the areas of the world that need lightning protection the most (Chap. 11) are the ones least likely to be able to afford it. What do we tell people in the rural or other unprotected areas to do and how do we make lightning protection affordable? Some ideas include:

1. Investigate alternative materials that can be sourced locally instead of importing European manufactured materials, saving both import and shipping fees, sometimes exorbitant prices and value-added tax (VAT). This includes recruiting recognized experts to design code-compliant LP with these materials.
2. Train local engineers and installers to install the systems instead of using expensive or untrained contractors.
3. Test and certify trained people so that the public can expect quality and code compliance.
4. Train and use local parents and students for the labor-intensive portion of an LP installation on schools, such as digging trenches around the buildings for laying the grounding circle, as well as discussing the principles of LP with those who are interested.
5. Collaborate with agencies from other countries in fund-raising to provide lightning protection for schools and other community buildings.

Questions

What other ideas could be used and tested to decrease cost but still provide code-compliant lightning protection?

Fig. 17.6 "Lightning eliminators" – not accepted by any lightning standard. (Courtesy of HCFP Pte. Ltd.)

References

Becerera M, Cooray V (2007) The early streamer emission principle does not work under natural lightning!! In: Preprints of the IX international symposium on lightning protection, Foz do Iguacu. 26–20 November 2007

Beccera M, Cooray V (2008) Laboratory experiments cannot be utilized to justify the action of early streamer emission terminals. J Phys D Appl Phys 41:8

Cooray V (ed) (2010) Lightning protection. IET, London

Federal Aviation Administration (1990) FAA standard: lightning protection grounding, bonding, and shielding requirements for facilities. https://www.faa.gov/about/office_org/headquarters_offices/ato/service_units/acquisition/grants/doc/FAA-12-01.pdf

Golde RH (1973) Lightning protection. Chemical Publishing, New York

Gomes C (2010) Modifications proposed to the concepts of structural protection. In: Preprints of the 30th international conference on lightning protection, 2010, Sardinia, 12–16 September 2010

Gomes C, Ab Kadir MZA (2011) Kadir Lightning protection: getting it wrong. IEEE Technology and Society Magazine 30(2):2–21

Gomes C, Ab Kadir MZA (2015) Efficient lightning protection: Optimization of economic, environmental and safety aspects. Environmental Engineering and Management Journal 14(8):1975–1985

Hartono ZA, Robiah I (2006) A review of studies on early streamer emission and charge transfer system conducted in Malaysia. In: Preprints of the 17th international symposium on electromagnetic compatibility, Zurich. 28 February-3 March 2006

Hartono ZA, Robiah I (2007) Improvised lightning protection system for indigenous homes and makeshift huts. Invited Lecture, 4th Asia lightning protection forum (ALPF), Guangzhou, China, February 2007

International Electrotechnical Commision (2010) IEC codes 62305–1,2,3,4

Kumarasinghe N (2008) A low cost lightning protection system and its effectiveness. In: Preprints of the international lightning meteorology conference. Vaisala, Tucson. 24–25 April 2008

Rakov V (2000) Lightning protection of structures and personal safety. In: Preprints of the international lightning detection conference. Vaisala, Tucson. 7–8 November 2008

Rakov VA, Uman MA (2003) Lightning: physics and effects. Cambridge University Press, New York

Uman M (2008) The art and science of lightning protection. Cambridge University Press, New York

Uman MA, Rakov VA (2002) A critical review of nonconventional approaches to lightning protection. Bull Am Meteorol Soc 98:1809–1820

Chapter 18
How to Build a Lightning Injury Prevention Program

Abstract It is far more preferable to prevent lightning injuries than to take care of the survivors and desolate families after they occur. This chapter will discuss factors that contribute to the effectiveness and reach of lightning injury prevention programs, provide an overview of most known lightning injury prevention programs internationally, review the effects of these programs, and discuss how to build an effective lightning injury prevention program. It will also provide cameos of a few individuals or groups who have devoted time and energy to investigate injuries and promote injury prevention. This chapter will not discuss lightning protection of nonliving things, which is covered in Chap. 17.

18.1 Philosophy Behind Lightning Injury Prevention

Some cultures believe that lightning injury is a curse from nature or from God and is inevitable, so they make little effort to avoid it. Others have tried to find ways to avoid lightning injury and death. In some histories and cultures, this took the form of propitiations to gods, in other totems, charms and potions to ward off spirits, or other formats. Many myths and practices based on cultural beliefs remain a challenge for educators and are discussed in Chaps. 10 and 21.

Although it is desirable to continue searching for treatment for those who have been injured by lightning, it is unlikely that the cascade of injury that lightning sets in motion will be easily stopped or reversed (Chaps. 2 and 4). Due to lack of funding, difficulty collecting enough patients to conduct controlled studies, and other factors, many lightning professionals have decided it is more productive to devote their time to prevention than to finding medical "cures."

18.2 How Do These Programs Start?

Lightning injury prevention programs often start when an individual or a community becomes upset about a lightning occurrence or about the overall number of injuries from lightning over a period of time and decides to do something about it.

M. A. Cooper, R. L. Holle, *Reducing Lightning Injuries Worldwide*,
Springer Natural Hazards, https://doi.org/10.1007/978-3-319-77563-0_18

They may be upset by a single event where a large number of people are injured or become aware of clusters of individuals or small groups injured. The list in Table 18.1 includes lightning physicists and electrical engineer researchers (Ralph Anderson, Chandima Gomes, Nobu Kitagawa, Shri Ram Sharma, and others), meteorologists (Ronald Holle, John Jensenius, Lyn Mainwaring, William Roeder, and others), and physicians (Chris Andrews, Mary Ann Cooper, Henk Jen ten Duis, Elizabeth Gourbière, Norberto Navarrete-Aldana, Ryan Blumenthal). Other programs have been initiated by government officials (Richard Tushemereirwe), military personnel (Sgt. Carlos Miguel Farias Malagon), environmentalists, and biologists (Munir Ahmed). Interestingly, no program known to these authors has been started by public health experts.

Sometimes these individuals were part of an existing research team. More often, they were individuals who, finding that friends and colleagues in their own fields did not always share their concern about lightning injuries, worked in isolation for many years. In time, especially with the dawn of electronic communication, many were able to find colleagues, often from other fields and sometimes in other countries, to work with them. Some found that multidisciplinary groups lent different views and expertise in solving research questions or contributed to broadening the scope of injury prevention education they could give for lightning safety.

The best early example of this approach (long before electronic communications) was Nobu Kitagawa, a lightning researcher in Japan, who recruited a physiologist and a physician (Drs. Ishikawa and Ohashi) to investigate aspects of casualty situations by devising often ingenious lightning laboratory investigations (Chap. 4). Kitagawa's group initiated the following studies:

- Investigated the lethal dose/kg of lightning using rodents
- Determined that death was a result of one maximum "dose" instead of a cumulative effect of the original lightning strike and return strokes (Ishikawa et al. 1985)
- Determined that survival depended on how rapidly flashover occurred (Ohashi et al. 1986)
- Investigated and replicated the concussive effects of lightning caused vapor explosions of surface moisture (Ohashi et al. 2001)
- Many other aspects of injury

After working on the problem in the laboratory, they would return to analyze the field evidence, applying what they had learned and modifying as appropriate (Kitagawa et al. 2001), finally, formulating lightning safety guidelines (Kitagawa et al. 1990; Andrews et al. 1996) that they hoped would be widely circulated to save lives. Dr. Kitagawa also mentored and advised others investigating lightning injuries in the laboratory (Andrews et al. 1989; Cooper and Kotsos 1997). Also see Chap. 4.

The largest and most effective recent program has been the Lightning Safety Group at the US National Oceanic and Atmospheric Administration (NOAA) that started in 2000, led by John Jensenius (Cooper and Holle 2012). Since this team

Table 18.1 Lightning[a] injury prevention programs and individuals by continent as of 2017

Africa
Kenya
ACLE-Kenya – Emerging program at Masinde Muliro University of Science and Technology (MMUST)
Carolyn Mulinya, Department of Geography and Earth Sciences
Malawi (Mulder et al. 2012)
ACLE-Malawi – New program at Malawi University of Science and Technology (MUST)
Leonard Kalindekafe, School of Climate and Earth Sciences
South Africa
Lightning Interest Group for Health, Technology and Science (LIGHTS) http://www.lightningsa.org.za
Ryan Blumenthal, forensic pathologist, University of Pretoria, (2006)
Ian Jandrell, Dean of Engineering, University of Witwatersrand (2009)
Ian McKechnie, engineer
Estelle Trengove, Head of Electrical Engineering, University of Witwatersrand
Ralph Anderson[a], A.E. Carte[a], engineers
Hugh Hunt, Engineering, University of Witwatersrand
Ken Nixon, Engineering, University of Witwatersrand Richard Evert, engineer, (2005)
Uganda (Ahurra 2012)
African Centres for Lightning and Electromagnetics Network (ACLENet) https://ACLENet.org
Richard Tushemereirwe, educator and government science officer
Mary Ann Cooper, physician
Ronald L. Holle, meteorologist
Jean Blaise Ngamini, meteorologist
Edmund Ataremwa, administrator
Zambia
ACLE-Zambia
Foster Chileshe Lubasi, engineer Zimbabwe (van Olst et al. 1990)
Asia
Bangladesh
Munir Ahmed, biologist
Ashraf Dewan, geographer
China (Zhang et al. 2010) India (Illiyas et al. 2014)
Lightning Awareness and Research Centre (LARC) https://lightningindia.wordpress.com/tag/larc/
V. Sasi Kumar
Chandima Gomes, engineer
Japan
Tomore Ishikawa
Nobu Kitagawa[a], Masajiro Ohashi[a]
Malaysia
Centre for Electromagnetic and Lightning Protection (CELP) http://www.eng.upm.edu.my/research/research_centres/centre_for_electromagnetic_lightning_protection_celp-2279

(continued)

Table 18.1 (continued)

Chandima Gomes, engineer, University of Putra Malaysia (2006)
Mohd Zainal Abidin ab Kadir, engineer, University of Putra Malaysia (2010, 2012)
Nepal
Shri Ram Sharma, physicist, Tribhuvan University
Singapore (Pakiam et al. 1981) Sri Lanka (Jayaratne et al. 2012)
Chandima Gomes, engineer
Australia
Christopher Andrews, physician, engineer
Europe
France
Gerard Berger
Nicholas Floret
Elizabeth Gourbière[a], physician
Netherlands
Henk Jen ten Duis, physician
Mexico (Raga et al. 2015) United Kingdom
Derek Elsom, Oxford Brookes University
North America
Canada
https://www.ec.gc.ca/foudre-lightning/
Lyn Mainwaring, meteorologist, ECCC
United States
Michael Cherington, physician
Mary Ann Cooper, physician (2010)
Donna Franklin, meteorologist, National Weather Service
Ronald Holle, meteorologist
John Gookin, National Outdoor Leadership School
Steve Hodanish, meteorologist, National Weather Service
John Jensenius, meteorologist, National Weather Service (2014)
Ken Langford, photographer
William Roeder, meteorologist Katie Walsh Flanagan (2013)
Philip Yarnell, physician
South America
Brazil (Cardoso et al. 2014) Colombia
Carlos Miguel Farias Malagon, military
Norberto Navarrete-Aldana, physician (2014)
Francisco Jose Roman Campos, engineer, Universidad Nacional de Colombia
Daniel Esteban Villamil Sierra, engineer

[a]Deceased

became active, lightning deaths in the United States have decreased from 55/year (10-year average from 1991 to 2000) to 27/year (10-year average from 2008 to 2017). This was accomplished through the use of public education, giving thousands of print and broadcast interviews on lightning injuries and safety information, providing lightning safety education tools to the 120 US National Weather Service offices, and teaching broadcast meteorologists' phrases such as "When Thunder Roars, Go Indoors."

The Lightning Safety Group's website http://www.lightningsafety.noaa.gov/ is recognized as the most complete and informative public website on lightning science and safety in the world. All of the information, public safety messages, animations, posters, games, and other materials are free for download and use by anyone, with team members and developers willing to help lightning safety programs on other continents and countries customize materials to be more "African" or "Asian," as desired. Although there is evidence that the Lightning Safety Group's campaign has decreased deaths from lightning in the United States over the past 17 years, there has been no recorded nor published monitoring of efficacy by other programs (Jensenius 2014).

18.3 Factors That Increase Effectiveness of Lightning Injury Prevention Programs

There are many factors that can increase the chances of success of a lightning injury prevention program in your country:

1. *Find like-minded partners*: Some efforts have languished when the individual initiator has not been able to recruit a team of partners to provide advice, manpower, and encouragement. Without partners, individual activists may find their enthusiasm wavering when their time and energy are filled by job pressures, family issues, illness, and other more pressing factors of life. Importantly, partners can take up the slack when one or more of the team is pulled away and provides valuable emotional encouragement, perspective, and other sources of support.
2. *Build a collaborative interdisciplinary team*: Bringing together individuals from different fields helps to educate all team members about a broader range of ideas and facts. Team members bring different skills, amplify ideas, and provide directions for education, research and service as well as valuable editing, fine-tuning of writing, and contacts to affected groups (Lengyel et al. 2010). Few of these programs have all of the individuals located in the same geographic area. For early programs, the "outside expert" (at least to the audience) that different team members can provide for each other enhances the credibility of each of the program individuals.

(a) An early example of this was the Lightning Safety Group, an ad hoc group of recognized lightning experts who met at the American Meteorological Society Annual Meeting in 1998 (American Meteorological Society 2003; Holle et al. 1999; Zimmermann et al. 2002). They produced the first unified set of lightning safety recommendations for individuals, small and large groups (Chap. 19).
(b) A more recent example is the NOAA Lightning Safety Group discussed in Sect. 18.2.

3. *Funding*: Most programs have been started by one person or a small number of people volunteering their time. Some have found partial funding from their government employers (NOAA/US National Weather Service; Environment and Climate Change Canada), from lightning safety organizations (Lightning Protection Institute, United States) or grants (Bangladesh). Even small amounts of funding can allow for printing of brochures, posters, and educational materials for schools, coaches, and parents, purchase of inexpensive souvenirs to be passed out at safety events (Fig. 18.1), travel to meetings to introduce the program or to present program updates and research, film public

Fig. 18.1 Non-Aligned Movement Science and Technology lightning meeting in Kathmandu, Nepal, 2011. From left: Arun P. Kulshreshtha (India), Mary Ann Cooper (United States), Shri Ram Sharma (Nepal), and Chandima Gomes (Malaysia) wearing easily transported, lightweight, crushable foam lightning visors that say "When Thunder Roars, Go Indoors." This meeting was instrumental in bringing together the founders of the African Centres for Lightning and Electromagnetics Network (ACLENet.org)

service announcements, and other activities that can help the program thrive and grow and reach much larger audiences than an individual alone can accomplish.

4. *Government support*: In the United States, there were several individuals involved in lightning injury prevention when the National Weather Service called for program ideas. John Jensenius submitted a proposal that was supported by several of these individuals in 2000. Lightning Safety Week was launched in 2001 by what became the year-round Lightning Safety Group. During the years that this program had monetary support, public service announcements, school curricula, and the best lightning safety website and educational and media resources in the world have been developed. The Lightning Safety Group has equipped every 1 of the 120 US National Weather Service offices with materials, facts, speakers, and encouragement to sponsor local activities as well as work with survivors and media to raise safety awareness. A history of lightning safety activities and work, primarily in the United States, is available at Cooper and Holle (2012) and Jensenius (2014). Unfortunately, despite great success, NOAA has not funded this team for several years and functionally disbanded it in 2016.
5. *Media involvement*: Lightning and lightning injury has so many aspects that it can become an ideal story for the media to cover as they choose different aspects as a focus (Table 18.2). Individuals or groups making themselves available for interviews and other interactions with the media can have many benefits:

    (a) News reports served as the basis for population studies in the United States from 1900 until use of the Internet became widespread, as well as use of Google searches, and other mechanisms for capturing injury information. As the media in any country becomes more interested and alert to the danger of lightning, they are more likely to report lightning injuries and deaths. As more are reported every year and, hopefully, reported in greater and more reliable detail, the public becomes more aware. Injury statistics also become easier to collect.

Table 18.2 Lightning: the perfect media story

Media factors	Lightning aspect
Beauty	Dramatic natural phenomenon
Science	Unusual, interesting, or new findings
Medicine	Aspects of injury
Common	Sports, work, recreation
Tragedy	Death/injury
Hope	Recovery and life after injury
Media coverage	Can make a difference
Public education	Injury prevention

Adapted from Cooper and Holle (2004)

(b) A media report usually reaches far more people than an individual lightning safety advocate. Encouraging the reporter to include lightning safety information in their report will save lives. The authors of this book ALWAYS stress that the media report "will save lives," something that reporters and journalists do not hear very often, and gives greater value to their hard work. Broadcast meteorologists can have a large audience through radio, television, or the Internet. Just as with other reporters, they can reach far more people and may be more credible than a lone safety activist. Teaching them the safety phrases and explanations to incorporate into their reports is time well spent.

6. *Utilize crises*: Although it is unfortunate that it often takes a crisis to raise the perceived value of a program, it is nevertheless true. Lightning safety programs should be well prepared to step into a crisis as an expert. Whether it is the lightning death or injury of someone famous or a government official's family member, the tragedy of multiple injuries or other heart-wrenching events, a crisis is an excellent time to turn the spotlight to prevention, education, and safety efforts.
7. *Publish and affiliate with a university or organization*: An institution that values academic output and community service can serve as an excellent base. Independent programs may fail due to lack of time and energy among their members who work full time and have family commitments. However, the group is more likely to survive and prosper if at least some of the members find that their work, research, publication, or community service is valued by their employer. Even better is when at least some of their lightning safety time is accommodated, their presentations at meetings are financially supported, and their efforts counted toward promotion.
8. *Mentor and network*: Networking can broaden perspective and understanding. It can give support to lone individuals. It is the obligation of lightning safety experts to mentor young faculty, students, and others around the world, so that we build a broad network that will survive the individuals who are now at work.
9. *Reliability and stability of infrastructure*: When power and communication systems are reliable or their use is inexpensive, lightning safety messages are more likely to reach their target audiences.
10. *Support and quality of government meteorological services*: In many countries, the meteorological services may have minimal financial support. Their mandate may be limited to support of military operations, aviation, or other programs important to the government. Having a positive relationship with government meteorologists may improve the distribution of warnings and forecasting availability and other involvement in safety and education efforts.
11. *Teach at every opportunity*: Some potential topics are listed in Table 18.3.

Table 18.3 Potential components of teaching that lightning injuries can be avoided

1. Counter common myths about lightning.
2. Teach lightning science.
3. Teach that "lightning safety is NOT convenient."
4. Stress planning ahead to avoid dangerous situations and to have a safety plan.
5. In developed countries, teach the three mottos of lightning safety (Roeder et al. 2009, 2011, 2015):
(a) When thunder roars, go indoors.
(b) NO PLACE OUTSIDE IS SAFE when thunderstorms are in the area.
(c) Half an hour since thunder roars, now it's safe to go outdoors.
6. Stress that the only "safe" places when lightning is present are in substantial buildings and metal-topped vehicles as long as the person is not in contact with the conducting path.
7. More detailed rules and explanations are in Cooper et al. (1999, 2017), Holle et al. (1999), Zimmermann et al. (2002), and www.lightningsafety.noaa.gov, among others.

18.4 Factors Hampering Development or Effectiveness of Lightning Injury Prevention

While all of these limitations can be overcome, each of the following factors make lightning injury programs more difficult to grow:

1. *Isolation of the activist.*
2. *Lack of funding and lack of support* from employer, family, and/or colleagues.
3. *Number of languages/diversity of the population, literacy rate.* In the United States, it is easy to use catchy, rhyming slogans such as "When Thunder Roars, Go Indoors" or "No Place Outside Is Safe When Thunderstorms Are In The Area," which even small children can remember – as long as English is their first language (Table 18.2). These slogans are short and have been tested to be easily remembered in an emergency situation. However, they are unlikely to translate well, particularly in settings with multiple languages. In addition, idiomatic meanings may prevent tailoring accurate, short or even polite delivery of a message that can be easily remembered in an emergency. In communities with low literacy, pictograms, alarms, colored beach flags, sirens, or other mechanisms may be necessary for warnings of acute danger (Chap. 20).
4. *Lack of easy solutions:* In developed countries, safe, substantial housing is widely available as well as fully enclosed all-metal vehicles, the two most important "safe" locations for improving survival from lightning. Unfortunately, when these are not available, such as in sub-Saharan Africa where 90% of the housing is not "lightning safe," different approaches and messages must be used. Unfortunately, most of these solutions have not been developed, tested, or scientifically validated.

5. *Cultural beliefs:* While there are many common and reasonably harmless myths about lightning that are often easy to counter, deeply held beliefs are much harder to address. Some of these have been held for generations and may have grown to be part of their relationship with nature, necessitating charms, potions, and incantations to avert. These cultures and their beliefs must be treated with respect, while at the same time, the lightning safety program looks for ways to go around these beliefs. Trengove studied folk beliefs common in South Africa for her dissertation, determining which were merely a nuisance or dangerous to the individual practicing them versus which were dangerous to communities (Trengove and Jandrell 2010).
6. *Emphasis on negatives:* Lightning safety advice can tend to emphasize what not to do, rather than the correct action (Table 19.3). The result may be a series of confusing, issue-specific recommendations that are difficult to remember when the lightning threat arises. In addition, the authors' experience has shown that media and public authorities often feel compelled to insert their personal incorrect opinions and interpretations about lightning safety that may be based on no knowledge of the subject.

18.5 Groups to Involve

As noted in Sect. 18.4, a core group of individuals from many disciplines increases the chances for success of sending the correct message as well as for doing research. Likewise, the success of delivering the message can be broadened by using groups that are negatively affected by lightning or who recognize lightning injury as a problem, such as parents, sports teams, broadcast meteorologists, and others. These groups may vary according to the setting, culture, beliefs, communications systems, and many other factors. Table 18.4 lists some groups that may be involved.

18.6 Lightning Injury Prevention Programs Around the World

There are many individuals and groups involved in lightning injury prevention in various locations; some are more active than others. It can safely be said that none of these people are involved in lightning injury prevention to make a profit and that they perform these activities only for the satisfaction of perhaps averting deaths and disabling injuries, generally for people they will never meet. Most do this work on their own time, often funding what they can afford as well.

Table 18.4 Potential groups to involve in lightning safety efforts

Community organizations
Churches
Community centers
Rotary, Civitan, 4-H, Lions
Construction industry
Insurance
Lightning protection
Famous national and local leaders
Government agencies
Military
National
Regional
Urban
US National Weather Service/NOAA
Industrial groups
Aviation
Building managers
Mining
Utilities, oil, gas
Media
Online entries
Print media
Radio
Television
Parents
Safety groups and officials
Emergency managers
Police, fire, and disaster managers
Safety seminars
Storm chasers/spotters
Schools and teachers
Curriculum planning
Elementary schools
High schools
Meteorology classes
Science fairs, school projects
Sports
Scientific and professional societies
Broadcasters
Electrical engineers
Lightning physicists
Medical groups
Meteorologists

(continued)

Table 18.4 (continued)

Lightning researchers
Public health professionals
Universities
Sports and outdoor groups
Agriculture
Coaches
Scouts and guides
Park managers
National Park Service
Managers of tourism facilities
Famous sports figures as spokespeople
Managers of heritage sites
Survivors and their families

Cooper and Holle (2004, 2005)

Table 18.5 National Weather Service Lightning Safety Group, United States

www.lightningsafety.noaa.gov
The lightning safety group began in 2001. Its website contains lightning safety recommendations, resources for media and educators, up-to-date demographics of US lightning fatalities and injuries in the current year, and a summary for the last decade. This team has worked with the media, broadcast meteorologists, and National Weather Service personnel to give hundreds of interviews and programs on lightning safety. United States lightning deaths have continued to decrease since the start of this team's activities. This is the most active lightning injury prevention program and the most complete website devoted to lightning safety education. Unfortunately, it has not received funding for several years and was functionally been disbanded by NOAA in 2016. Several of the key members have been recruited to the National Lightning Safety Council as a private successor group.
Focal point: John Jensenius, National Weather Service, Gray, Maine.

Table 18.6 African Centres for Lightning and Electromagnetics Network, Inc. (ACLENet)

ACLENet.org
ACLENet began in 2013 and is a pan-African network of national and regional centers dedicated to decreasing deaths, injuries, and property damage from lightning. Its objectives include developing public education on lightning safety that is appropriate for Africa, working with governments to address lightning safety and protection, graduate training to produce Africa's own experts, lightning protection of infrastructure, and protection of schools. This is the first such program in Africa.
National centers have been formed in Zambia (ACLE-Zambia), Kenya (ACLE-Kenya), and Malawi (ACLE-Malawi) with other countries expressing interest.
Focal point: Mary Ann Cooper, MD, director.

If you are interested in doing research on injuries in your country or in doing injury prevention, we hope you will reach out to some in the lists in Tables 18.1, 18.5, and 18.6. We hope you will find, as we have, that lightning injury prevention people are some of the nicest and easiest people to work with that you will encounter in your professional life. You may also find that some, depending on their expertise and time, may offer to edit, cross-check, and enhance your materials, help you plan research, help you to modify their safety materials for your country, give emotional support and expertise, or even become a mentor-colleague and sometimes a coauthor on research you may decide to pursue together. References at the end of this chapter indicate some of the publications that have been produced by these individuals and groups; their geographic areas and topics are evident from the publications' titles. Chapter 6 also has a review of recent national summaries of lightning death and injury statistics.

References

Ab Kadir MZA, Cooper MA, Gomes C (2010) An overview of the global statistics on lightning fatalities. In: Preprints of the 30th international conference on lightning protection, Cagliari. 13–17 September 2010

Ab Kadir MZA, Misbah NR, Gomes C et al (2012) Recent statistics on lightning fatalities in Malaysia. In: Preprints of the 31st international conference on lightning protection, Vienna. 2–7 September 2012

Ahurra M, Gomes C (2012) Lightning injuries in Uganda. In: Preprints of the 31st international conference on lightning protection, Vienna. 2–7 September 2012

American Meteorological Society (2003) Lightning safety awareness [AMS statement and background from AMS Council]. Bull Am Meteorol Soc 84:260–261

Andrews CJ, Berger G, Floret N et al (1996) International safety and rescue guide against lightning hazards. In: Proceedings of the 23rd international conference on lightning protection, Florence. 23–27 September 1996

Andrews CJ, Darveniza M, Mackerras D (1989) Lightning injury: a review of clinical aspects, pathophysiology and treatment. Adv Trauma 4:241–287

Bennett BL, Holle FL, Cooper MA (2006) Lightning safety. In: Clossner D (ed) 2006-07 NCAA sports medicine handbook, 18th edn. National Collegiate Athletic Association, Indianapolis

Blumenthal R (2006) When thunder roars -- go indoors! S Afr Med J 96(1):38–39

Cardoso I, Pinto O Jr, Pinto IRCA et al (2014) Lightning casualty demographics in Brazil and their implications for safety rules. Atmos Res 135-136:374–379

Cooper MA, Holle RL (2004) How to mobilize the information systems in your country to change lightning safety standards (and save lives and injuries). In: Preprints of the 18th international lightning detection conference. Vaisala, Helsinki. 7–9 June 2004

Cooper MA, Holle RL (2005) How to use public education to change lightning safety standards (and save lives and injuries). In: Preprints of the 14th symposium on education. American Meteorological Society, San Diego. 9–13 January 2005

Cooper MA, Holle RL (2010) Mechanisms of lightning injury should affect lightning safety messages. In: Preprints of the 3rd international lightning meteorology conference, Vaisala, Orlando. 21–22 April 2010

Cooper MA, Holle RL (2012) Lightning safety campaigns - USA experience. In: Preprints of the 31st international conference on lightning protection, Vienna. 2–7 September 2012

Cooper MA, Andrews CJ, Holle RL et al (2017) Lightning-related injuries and safety. In: Auerbach (ed) Wilderness medicine, 7th edn. Elsevier, Philadelphia, pp 71–117

Cooper MA, Kotsos TP (1997) Development of an animal model of lightning injury with flashover utilizing a table-top lightning generator. In: Proceedings of lightning and mountains '97, Chamonix Mont-Blanc. 1–5 June 1997

Cooper MA, Holle R, López R (1999) Recommendations for lightning safety. J Am Med Assoc 282:1132–1133

Evert R, Schulze G (2005) Impact of a new lightning detection and location system in South Africa. In: Proceedings of the inaugural IEEE PES 2005 conference and exposition in Africa, Durban. 11–15 July 2005

Gomes C, Kithil R, Ahmed M (2006) Developing a lightning awareness program model for third world based on American-south Asian experience. In: Preprints of the 28th international conference on lightning protection, Kanazawa. 18–22 September 2006

Holle RL, Lopez RE, Zimmermann C (1999) Updated recommendations for lightning safety - 1998. Bull Am Meteorol Soc 80(10):2035–2041

Illiyas FT, Mohan K, Mani SK et al (2014) Lightning risk in India: challenges in disaster compensation. Economic & Political Weekly, 7 June, XLIX:23-27

Ishikawa T, Ohashi M, Kitagawa N et al (1985) Experimental study on the lethal threshold value of multiple successive voltage impulses to rabbits simulating multi-strike lightning flash. Int J Biometeorol 29(2):157–168

Jandrell I, Blumenthal R, Anderson R et al (2009) Opening session and keynote address: recent lightning research in South Africa with a special focus on keraunopathology. In: Proceedings of the 16th international symposium on high voltage engineering, Cape Town. 24–28 August 2009

Jayaratne K, Gomes C (2012) Public perceptions and lightning safety education in Sri Lanka. In: Preprints of the 31st international conference on lightning protection, Vienna. 2–7 September 2012

Jensenius J, Franklin DB (2014) NOAA's efforts to reduce lightning fatalities through public education and awareness. In: Preprints of the 5th international lightning meteorology conference. Vaisala, Tucson. 20–21 March 2014

Kitagawa N, Ohashi M, Ishikawa T (1990) Safety guide against lightning hazards. Res Lett Atmos Electr 10:37–44

Kitagawa N, Ohashi M, Ishikawa T (2001) The substantial mechanisms of step voltage effects. Atmos Electr 21(2):87–94

Lengyel MM, Cooper MA, Holle RL et al (2010) The role of multidisciplinary teams and public education in decreasing lightning casualties worldwide. In: Preprints of the 30th international conference on lightning protection, Cagliari. 13–17 September 2010

Mulder MB, Msalu L, Caro T et al (2012) Remarkable rates of lightning strike mortality in Malawi. PLoS One 7(1)

Navarrete-Aldana N, Cooper MA, Holle RL (2014) Lightning fatalities in Colombia from 2000 to 2009. Nat Hazards 74:1349–1362

Ohashi M, Kitagawa N, Ishikawa T (1986) Lightning injury caused by discharges accompanying flashovers--a clinical and experimental study of death and survival. Burns Incl Therm Inj 12(7):496–501

Ohashi M, Hosoda Y, Fujishiro Y et al (2001) Lightning injury as a blast injury of skull, brain, and visceral lesions: clinical and experimental evidences. Keio J Med 50:257–262

Pakiam J, Chao T, Chia J (1981) Lightning fatalities in Singapore. Meteorol Mag 110:175–187

Raga G, de la Parra MG, Kucienska B (2015) Deaths by lightning in Mexico (1997-2011): threat or vulnerability. Weather Clim Soc 6:434-444

Roeder WP (2009) Research required to improve lightning safety. In: Preprints of the 4th conference on the meteorological applications of lightning data. American Meteorological Society, Phoenix. 11–15 January 2009

Roeder WP, Holle RL, Cooper MA et al (2011) Communicating lightning safety effectively. In: Preprints of the 5th conference on the meteorological applications of lightning data. American Meteorological Society, Seattle. 23–27 January 2011

Roeder WP, Cummins BH, Cummins KL et al (2015) Lightning fatality risk map of the contiguous United States. Nat Hazards 79:1681–1692

Trengove E, Jandrell IR (2010) Strategies for understanding lightning myths and beliefs. In: Preprints of the 30th international conference on lightning protection, Cagliari. 13–17 September 2010

van Olst MDA, (1990) Minimising lightning fatalities: lightning earth currents in Zimbabwe. In Procceedings of the 1st all Africa international symposuim on lightning, Harare30 April-4 May 1990

Walsh KM, Bennett B, Cooper MA et al (2013) National Athletic Trainers' association position statement: lightning safety for athletics and recreation (see also errata in 38(1):83). J Athl Train 48(2):258–270

Zhang W, Meng Q, Ma M et al (2010) Lightning casualties and damages in China from 1997 to 2009. Nat Hazards. https://doi.org/10.1007/s11069-010-9628-0

Zimmermann C, Cooper MA, Holle RL (2002) Lightning safety guidelines. Ann Emerg Med 39(6):660–665

Chapter 19
Lightning Safety Guidelines and Resources

Abstract The earliest lightning safety recommendations were often based on untested myths, superstitions, religious beliefs, and other folklore. This chapter will give a history of how modern lightning safety guidelines were formed. Prior to 1990, lightning safety advice had not been updated for several decades. Renewed interest in the United States was due to several factors including medical studies, recognition of the lightning impacts on athletics, availability of real-time lightning detection data, analysis of mechanisms of injury, and casualty data collection for large sample sizes. Recent guideline modifications now stress lightning "safe" areas as substantial buildings and metal vehicles and are clear that complete safety cannot be achieved outside these two locations. Unfortunately, most modern lightning safety recommendations apply only to developed countries, where "safe locations" are nearly always available. Developing countries, where there are no or very few "safe locations," are in desperate need of guidelines that are useful, valid, inexpensive, and easily implemented.

19.1 US Safety Material from the 1950s to the 1990s

The US Weather Bureau (now National Weather Service) publication, *Storm Data*, began in the 1950s and continues to the present. It is among the best and longest-duration sources of weather-related impacts for any nation in the world. It includes data on the impacts of all types of weather, including thunderstorms and lightning, tornadoes, hail, heat, cold, snow, floods, hurricanes, ice storms, and strong winds.

When *Storm Data* began in the 1950s, lightning incident location categories were established when Weather Bureau staff apparently reviewed a limited number of lightning cases that were available at the time when *Storm Data* was established and developed the code of 0 through 9 in Table 19.1. Curran et al. (2000) summarized these categories for the period of 1959 to 1994. Symptomatic of the understanding of lightning casualties at that time is the emphasis on golf and the vague term "open fields."

Ideally, injury categories should be helpful in developing prevention, warning, and safety guidelines. Unfortunately, the categories in Table 19.1 were skewed and uninformative. For instance, unknown (9) and open fields (6) accounted for more

M. A. Cooper, R. L. Holle, *Reducing Lightning Injuries Worldwide*,
Springer Natural Hazards, https://doi.org/10.1007/978-3-319-77563-0_19

Table 19.1 Locations of lightning casualties in *Storm Data* including unknown cases in the United States from 1959 through 1994

Code	Location of casualty	Percent of total (%)
1	Under trees	14
2	Water related, fishing, boating, swimming, etc.	8
3	Golfing	4
4	Golfing and under trees	1
5	Driving tractors, farm equipment, heavy road equipment, etc.	3
6	Open fields, ballparks, playgrounds, etc.	27
7	Telephone related	2
8	Radios, transmitters, antennas, etc.	1
0, 9	Not reported, at various other and unknown locations	40

Curran et al. (2000)

than half of all cases, while most other categories were infrequent. Injury prevention recommendations for those in "open fields" might be very different depending on whether the incident occurred in an open field to someone performing labor-intensive subsistence agriculture compared with someone on a recreational hike across an open meadow in the mountains during a summer day. Noting both the location and activity for each case, a technique used since Holle et al. (2005), is a much more informative and useful method for injury prevention.

19.2 Safety Material Starting in the 1990s

There was resistance in the United States to revising prior safety materials, and many of the other early lightning safety recommendations were at odds with one another (Uman 1986; Vavrek et al. 1993; Cooper and Holle 2005). One of the earliest international safety guides to be proposed and based on scientific data was by Ishikawa, Kitagawa, and associates (Kitagawa et al. 1990; Andrews et al. 1996, 1997). Unfortunately, these did not enjoy wide distribution, so had little measurable impact.

Fortunately, in the 1990s, a number of major factors brought into question the assumptions that resulted in the *Storm Data* categories. Medical examinations of casualties were showing that the direct strike was not occurring as often as had been thought, while ground current and other mechanisms were much more frequent (Chap. 2) (Andrews et al. 1992; Cooper et al. 2008). The formation of the support group Lightning Strike and Electric Shock Survivors, International, during this period also presented lightning casualties differently than had been expected (Cooper and Marshburn 2005).

During this time, real-time lightning detection networks were being brought into routine operation and showed what was occurring in the vicinity of lightning casualties. The results were not as expected (Holle et al. 1993). Critical observations from this study over Florida in the United States revealed that six people were killed or injured by lightning on a beach from a storm containing only two cloud-to-ground flashes. Additional central Florida cases showed that casualties occurred before, during, and after the most frequent lightning in a thunderstorm. Also, casualties were spread across weak, moderate, and strong thunderstorms. These results were subsequently affirmed in demographic and case studies by Lengyel et al. (2005), Hodanish and Zajac (2002), and Hodanish et al. (2004, 2015). The ability of such studies to composite well-located lightning relative to deaths and injuries led to major reexamination of lightning safety advice to stress the lightning danger of all phases and intensities of thunderstorms.

Special attention and credit must be given to the National Collegiate Athletic Association (NCAA) who initiated lightning safety in the late 1990s when research found that almost no major US colleges had lightning safety plans for either their sports practices or for fans at stadium games (Walsh and Bennett 1996; Bennett 1997; Bennett et al. 1997). The NCAA's unilateral decision to include lightning safety in the *NCAA Sports Medicine Handbook* was instrumental in changing lightning safety standards for sports at all levels and ages. It has been updated several times (Bennett et al. 1997, 2006; subsequent versions without attributed authorship). Similar recommendations had been initiated within the sports medicine community by Bennett (1997), Walsh (1997), and Walsh et al. (1997). The most complete recent athletics lightning safety policy was developed and published by the National Athletic Trainers' Association (NATA; Walsh et al. 2013). The participation of NATA is appropriate since athletic trainers may be more likely to watch for the development of lightning than coaches as well as to treat those who are casualties.

The issues brought up by the pioneering medical, athletics, meteorological, and demographic studies were mostly limited to a few lightning and related communities in the 1990s. Each approach attracted people who took a fresh look at the data, suggesting revisions to the old safety materials from the 1950s to the1990s. Research on lightning inter-strike distances finally spurred individuals to meet; formulate a consensus statement of recommendations on lightning risk, safety, and injury prevention (López and Holle 1999); and coordinate the dissemination of these safety messages, especially to the media. Recognized lightning experts from many fields including meteorology, education, engineering, research, academics, medicine, physics, insurance, sports, and the lightning protection industry gathered for an ad hoc meeting at the American Meteorological Society's Annual Meeting in 1998 (Table 19.2). Most had been involved in individual lightning safety briefings, media interviews, and publications, although very few were fully engaged in lightning safety policies and studies. Despite crossing disciplines, most already knew and valued the others as professional colleagues.

Several of this group's ideas were modified from the pioneering National College Athletic Association lightning safety guidelines. The lightning safety guidelines formulated at this meeting included safety for individuals, for small groups with short evacuation times, and for large groups with longer evacuation times such as sports stadia or rock concert venues. They also discussed how to

Table 19.2 Members of the Lightning Safety Group who met at the 1998 American Meteorological Society conference and others whose opinions were sought

Name	Discipline	Affiliation	Country
C. Andrews[a]	Physician	Medical center	Australia
B. Bennett	Athletic trainer	University	United States
L. Byerley	Lightning protection engineer	Business	United States
M. Cherington[a]	Physician	Hospital	United States
M.A. Cooper	Physician	University	United States
K.L. Cummins	Lightning detection engineer	Business	United States
E. Gourbière[a]	Physician	Federal laboratory	France
G. Harwood[a]	Writer	Business	United States
R.L. Holle	Research meteorologist	Federal laboratory	United States
K.W. Howard	Research meteorologist	Federal laboratory	United States
R. Kithil	Lightning protection engineer	Business	United States
E.P. Krider	Professor of physics	University	United States
L.C. Lawry	Lightning detection manager	Business	United States
R.E. López	Research meteorologist	Federal laboratory	United States
B. Lunning	Loss control specialist	Business	United States
J.T. Madura	Manager	Rocket launch facility	United States
M. McGee	Lightning protection engineer	Business	United States
C. Ojala[a]	Professor	University	United States
M. Primeau[a]	Neuropsychologist	University	United States
W.P. Roeder	Meteorologist	Military	United States
J. Vavrek	Science teacher	Middle School	United States
K. Walsh Flanagan[a]	Athletic trainer	University	United States
C. Zimmerman	Safety management	Business	United States

Zimmermann et al. (2002)
[a]Not at meeting

formulate a lightning safety action plan and the "30–30 rule." Recommendations noted that the only safer places to seek when lightning threatened were inside a substantial building containing indoor plumbing, wiring, and framing in the walls or a fully enclosed all-metal vehicle. Danger was emphasized before and after thunderstorms as well as at their most intense stage. These initial lightning safety guidelines were published as widely as possible in the respected journals of the participants (Holle et al. 1999; Zimmermann et al. 2002; American Meteorological Society 2003).

An important conclusion of the group was that lightning safety and education is an international, interdisciplinary endeavor. Before publication, the proposed guidelines were shared for review and comment with many respected lightning colleagues from around the world who were unable to attend the meeting personally (Table 19.2). An additional lesson learned over the years since the 1998 American Meteorological Society meeting was that lightning safety efforts in other countries are much more effective if such a diverse group with independent interests in lightning work together and exchange ideas and information (Chap. 18).

19.3 Modifications and Additions to Lightning Safety Guidelines

As a result of this meeting and studies by its participants, the US National Weather Service began an effort to continue communication among those who were involved in the 1998 American Meteorological Society meeting. One outcome was monthly conference calls during the summer. Another was to post the updated recommendations on the emerging website technology at www.lightningsafety.noaa.gov. The Lightning Safety Group was formed and extensive education was done of National Weather Service personnel as well as broadcast meteorologists, especially during Lightning Safety Week, adopted as the last full week of June. The recommendations have been incorporated into the literature for many sports publications, magazines, and coaches. Recently dedicating 1 week for lightning safety has been abandoned in favor of teaching it when needed in each part of the country (Table 19.3).

A similar effort in Canada resulted in new materials such as videos and simulations in both the English and French languages (Mainwaring and Fricska 2016). In addition, Environment and Climate Change Canada has a free website showing real-time lightning data every 10 min at http://weather.gc.ca/lightning/index_e.html.

The identification of the five mechanisms of lightning injury clarified the risk by minimizing the direct-strike hypothesis and raising awareness of ground current/step voltage and other mechanisms (Chap. 2; Cooper et al. 2008). The two largest public perceptions that still need to be changed in the developed world are the over-reliance on avoiding the direct strike and the impossibility of attaining certain lightning safety while outdoors. While there has indeed been substantial modification of the prior material in the United States and Canada, for example, experience showed that changes in safety advice take years or decades to permeate all aspects of available materials (Table 19.3).

A difficult group to educate about lightning safety is the outdoor recreation population who consciously choose to be away from the safety of lightning-safe buildings and vehicles for their activities but still wish for fail-safe, "lightning safety" recommendations. Unfortunately, their activities also often coincide with peak lightning times, such as summer afternoons, and locations, such as mountainous terrain or water activities. The National Outdoor Leadership School (NOLS) has developed educational materials with an honest assessment of the lightning recommendations and risks in such situations (Gookin and Morris 2014). In every step of the recommendations, NOLS makes it clear that "no place outdoors is safe from lightning" (Table 19.4). This phrase is repeated often, strongly advising outdoor recreationalists that, short of rescheduling their activity or having the sure availability of a lightning-safe evacuation area, any other precautions will not insure any measurable degree of safety.

Methods for safety and warning have evolved over time (Table 19.3). Some have been discarded and others have been changed or updated based on research

Table 19.3 Warning and safety methods that have been modified or abandoned since development of Lightning Safety Group

Method	Application	Reasons for change
Flash-to-bang	Count the seconds from seeing lightning to hearing thunder, then divide by 5 (miles) or 3 (km) to find the distance to lightning	1. People do not easily remember to divide by 5 or 3, resulting in an underestimate of the distance to lightning
		2. It can be difficult to match the sound with the corresponding lightning
Don't lists	Extensive lists of what NOT to do, such as:	1. The DO message directs a specific action, rather than hoping that people will decide what to do on their own from a long list of Don'ts
	Don't stand under trees	
	Don't be by water	
	Don't be tallest object in a clearing	2. Simple, short, specific DO messages are better remembered than a long series of confusing issue-specific rules
	Don't touch metal or plumbing	
	Don't do this, don't do that, etc.	
Crouch	The method is to crouch down to decrease height, with feet close together to decrease ground current footprint, and hands over the ears to decrease the chance of eardrum rupture	1. The method applies only to direct strikes, which are a mere 3–5% of deaths.
		2. Roeder (2014) showed that the crouch improves chances of survival by only 53%. The result is a paltry 2–3% improvement in fatality risk
		3. Roeder (2009, 2014) found that standing with feet together is equally as effective as the crouch, which is difficult to position and maintain for any length of time
30–30 rule	The first 30 indicates that evacuation to a safe area should occur if the time from seeing lightning to hearing thunder is 30 s or less. The second 30 refers to how many minutes should elapse before returning to outdoor activity after last lightning is seen or thunder is heard	1. Counting 30 before evacuating wastes time that could be used to reach safety
		2. The first 30 in seconds is for a six-mile distance to the nearest lightning. It is used in commercial applications and often adjusted to five miles
		3. The second 30 in minutes is a standard for large crowd control.
		4. Applying 30–30 in personal situations can be confusing.
Lightning safety week	Last full week of June	No single week is the best time for safety week across a large region due to known variations in the time of year when lightning occurs (Chap. 11). Safety should be taught and practiced throughout the year

Table 19.4 Mottos currently in use for lightning safety in the United States

When thunder roars, go indoors
No place outside is safe when thunderstorms are in the area
30 min since thunder roars, now it's safe to go outdoors

and "best practice" (Vavrek et al. 1993; Lengyel et al. 2005; Roeder 2007, 2009, 2014; Roeder et al. 2012; Cooper and Holle 2012; Jensenius et al. 2008). These have been incorporated into the information available on the Lightning Safety website www.lightningsafety.noaa.gov, which also has teaching resources, children's games, and media tools. Frequent updates to the website are made (Chap. 18) including the investigation and listing of each lightning death as it occurs in the United States. All of the materials on this website are free for download and printing by anyone in the world.

19.4 Lightning Safety Mottos and Toolkits

Several safety mottos that have been tested in focus groups and other situations in the United States are in Table 19.4. They are short and easy to remember, even by 3-year-old children, and have been taught to weather broadcast meteorologists across the nation as well as used by National Weather Service personnel in outreach and training programs. Unfortunately, these may not be easily translated to other languages. In sub-Saharan Africa and many other parts of the world where the predominant housing is still mud bricks with flammable thatch or sheet metal roofing, "When Thunder Roars, Go Indoors" may be the worst advice to give.

Another area of concern are large sports or entertainment venues such as sports stadia, car races, public beaches, amusement parks, music concerts, and horse racing. These events often have huge crowds over large areas, often with inadequate safety areas, signage, weather monitoring, or public warning systems. Thunder, the prime warning to individuals, may not be easily heard over the noise of the cars or crowds. Fortunately, leaders from the Lightning Safety Week committee have assembled lightning safety toolkits which have been adopted by many venues and organizations (Woodrum and Franklin 2012; http://www.lightningsafety.noaa.gov/toolkits.shtml).

Wilderness situations have been addressed where there are no "safe" areas for evacuation, such as fishing camps, white water rafting in canyons, and hiking wilderness trails (Gookin 2002, 2010). Wise outdoor recreationists should always recognize that there are almost always trade-offs in safety and availability of medical care in wilderness situations. Lightning, like all other wilderness and environmental risks, requires preplanning just as much as foreseeing water needs in a desert, bear avoidance in the mountains, or risk of drowning and hypothermia in water sports.

19.5 International Lightning Safety

Lightning safety guidelines and materials on the NOAA Lightning Safety website, (http://www.lightningsafety.noaa.gov) are all free for download and modification for use in other countries. Unfortunately, most of the recommendations apply well only to developed countries.

Lightning safety advocates have arisen in many countries (Chap. 18) but have often been lone voices with little real effect on lightning deaths and injuries. This is beginning to change. Over the last decade, four symposia on lightning have been organized by patronage of the Non-Aligned Movement Science and Technology Centre (http://www.namstct.org/nam_s&t_his.htm). These symposia have been in the following four countries: Sri Lanka in 2007, Nepal in 2011, Uganda in 2013, and Zambia in 2015.

At each conference, issues related to lightning safety in developing countries were discussed and recommendations documented and individual papers collected as proceedings (Arora and Gomes 2008; Sharma 2012; Trengove and Lubasi 2015; Holle and Ataremwa 2017). In addition, the following resolutions have been passed for action at various levels:

- Colombo Declaration
- Kathmandu Resolution
- Entebbe Resolution
- Resolution for Declaration of an International Lightning Safety Day (https://ACLENet.org/publications)

A substantive outcome of these meetings was the 2014 founding of the African Centres for Lightning and Electromagnetics Network (ACLENet), a pan-African network of national and regional centers dedicated to decreasing deaths, injuries, and property damage from lightning (Cooper et al. 2016). Nevertheless, ACLENet is struggling with finding valid lightning safety information to use for public education in communities where there are no safe places to reach when thunderstorms are in the area. While ACLENet's primary focus is Africa, the organization:

- Collaborates with lightning researchers and safety advocates worldwide
- Provides a focus for lightning safety advocacy
- Catalyzes research into appropriate lightning protection, safety practices, and recommendations for developing countries (Chap. 17)
- Promotes undergraduate, graduate, and professional training and improvement of engineering and meteorological services in Africa
- Advocates lightning protection of some of the most vulnerable, Africa's school children

19.6 Lightning Safety Materials

Lightning safety recommendations are listed on many websites and media articles, some of which were posted years ago and have not been updated to incorporate current practices. Unfortunately, readers may not be aware that they may be reading outdated material.

The National Weather Service site described in Table 19.5 serves as the current "gold standard" since it has the most up-to-date and recently reviewed guidelines and is monitored by expert lightning safety researchers and educators. All materials on this site are free for download. In addition, many of the source materials can be made available on request to the authors to enable tailoring the materials to show scenes and people from countries such as those in Africa or Asia, as well as being available for translation to other languages. An outgrowth of this site is the National Lightning Safety Council (http://www.lightningsafetycouncil.org/LSC-Home.html) that is comprised of the principal members of the Lightning Safety Team committee.

The context of the materials given in many US-based websites focuses on lightning safety issues related to the lifestyle of US public: ready accessibility to weather sites (electronic and audiovisual media), fast dissemination of risk awareness through modern communication modes, and ready availability of sturdy structures and metal vehicles. Additionally, the guidelines also discuss the safety concerns related to a variety of activities common in the US day-to-day life: indoor and outdoor swimming pools, leisure and adventure camping, golfing, and other sports/recreational activities.

While these may be applicable to many other developed countries, there is a severe shortage of lightning safety information and recommendations that can apply to developing countries. These sites are included in Table 19.6.

Table 19.5 US National Weather Service website for lightning safety

Name of group	Geographical area	Began
NWS Lightning Safety Team	United States	2001
Resources		
www.lightningsafety.noaa.gov		
Website contains lightning safety recommendations, resources for media and educators, games for children, and up-to-date demographics of US lightning fatalities and injuries in the current year and a summary for last decade.		
Focal point		
John Jensenius, National Weather Service, Gray, Maine		
Comments		
This is the most complete website devoted to lightning safety education. Everything is free for download. Safety educators from other countries may request source materials that they can modify with pictures, language, and idioms consistent with their own country.		

Table 19.6 Websites with lightning information in developing countries and other resources (accessed October 2017)

Country	Organization
Website	
Canada	ECCC
http://www.ec.gc.ca/foudre-lightning/default.asp?lang=En&n=159F8282-1	
Colombia	PREVER
https://sites.google.com/site/lesionesxrayos	
India	LARC
https://lightningindia.wordpress.com/	
Nepal	
https://www.youtube.com/watch?v=yWDgHNGNXAY	
South Africa	LIGHTS
http://www.lightningsa.org.za/	
United States	
www.lightningsafety.noaa.gov	NOAA
https://scijinks.gov/zap-game/	
https://vimeopro.com/asctraining/asc-safety-training/video/40944630	
Other	
https://www.dehn-international.com/sites/default/files/uploads/dehn/pdf/Kataloge/Englisch/Brochures/ds661_e_when_lightning_strikes.pdf	
https://www.lds.org/callings/church-safety-and-health/training-and-video-resources/lightning-safety?lang=eng	

References

American Meteorological Society (2003) Lightning safety awareness [Adopted by AMS Council on 29 April 2002]. Bull Am Meteorol Soc 84:260–261

Andrews CJ, Cooper MA, Darveniza M et al (1992) Lightning injuries: electrical, medical, and legal aspects. CRC Press, Boca Raton

Andrews CJ, Berger G, Floret N et al (1996) International safety and rescue guide against lightning hazards. In: Proceedings of the 23rd international conference on lightning protection, Florence, 23–27 Sept 1996

Andrews CJ, Berger G, Floret N et al (1997) International safety and rescue guide against lightning hazards. In: Proceedings of lightning and mountains '97, Chamonix Mont-Blanc, 1–5 June 1997

Arora R, Gomes C (2008) Need for awareness, detection and protection from damage caused by lightning. Centre for Science and Technology of the Non-Aligned and Other Developing Countries, New Delhi. ISBN 978-81-7035-579-3

Bennett BL (1997) A model lightning safety policy for athletics. J Athl Train 32:251–253

Bennett BL, Holle RL, López RE (1997) Lightning safety. In: Benson (ed) 1997–98 NCAA sports medicine handbook. National Collegiate Athletic Association, Overland Park

Bennett BL, Holle RL, Cooper MA (2006) Lightning safety. In: Clossner D (ed) 2006–07 NCAA sports medicine handbook, 18th edn. National Collegiate Athletic Association, Indianapolis

Cooper MA, Holle RL (2005) How to use public education to change lightning safety standards (and save lives and injuries). In: Preprints of the 14th symposium on education, American Meteorological Society, San Diego, 9–13 Jan

Cooper MA, Holle RL (2012) Lightning safety campaigns – USA experience. In: Preprints of the 31st international conference on lightning protection, Vienna, 2–7 Sept 2012

Cooper MA, Marshburn S (2005) Lightning strike and electric shock survivors, international. NeuroRehabilitation 20:43–47

Cooper MA, Holle RL, Andrews CJ (2008) Distribution of lightning injury mechanisms. In: Preprints of the 2nd international lightning meteorology conference, Vaisala, Tucson, 21–23 Apr 2008

Cooper MA, Gomes C, Tushemereirwe R et al (2016) The development of the African Centres for Lightning and Electromagnetics network. In: Preprints of the 33rd international conference on lightning protection, Estoril, 25–30 Sept 2016

Curran EB, Holle RL, López RE (2000) Lightning casualties and damages in the United States from 1959 to 1994. J Clim 13:3448–3453

Gookin J (2002) NOLS backcountry lightning safety guidelines. In: Preprints of the 17th international lightning detection conference, Vaisala, Tucson, 16–18 Oct 2002

Gookin J (2010) Backcountry lightning risk management. In: Preprints of the 3rd international lightning meteorology conference, Vaisala, Orlando, 21–22 Apr 2010

Gookin J, Morris S (2014) Lightning. National Outdoor Leadership School, Stackpole Books, Mechanicsburg

Hodanish S, Zajac B (2002) Documentation of the "First lightning flash of the day" associated with a weak shallow convective updraft killing an 18 year old on top of Pikes Peak, Colorado. In: Preprints of the 17th international lightning detection conference, Vaisala, Tucson, 16–18 Oct 2002

Hodanish S, Holle RL, Lindsey DT (2004) A small updraft producing a fatal lightning flash. Weather Forecast 19:627–632

Hodanish S, Wolyn P, Mozley K (2015) Meteorological analysis of the Rocky Mountain National Park lightning fatalities of 11 and 12 July, 2014. In: Preprints of the 7th conference on the meteorological applications of lightning data, American Meteorological Society, Phoenix, 4–8 Jan 2015

Holle RL, Ataremwa E (2017) Lightning impacts in developing countries of Africa and Asia. Centre for Science and Technology of the Non-Aligned and Other Developing Countries, New Delhi

Holle RL, López RE, Ortiz R et al (1993) The local meteorological environment of lightning casualties in central Florida. In: Preprints of the 17th conference on severe local storms and conference on atmospheric electricity, American Meteorological Society, St. Louis, 4–8 Oct 1993

Holle RL, Lopez RE, Zimmermann C (1999) Updated recommendations for lightning safety – 1998. Bull Am Meteorol Soc 80:2035–2041

Holle RL, López RE, Navarro BC (2005) Deaths, injuries, and damages from lightning in the United States in the 1890s in comparison with the 1990s. J Appl Meteorol 44:1563–1573

Jensenius JS, Franklin DB, Hodanish S (2008) Lightning kills – play it safe: NOAA's continuing efforts to educate the public on lightning and lightning safety. In: Preprints of the 2nd international meteorology conference, Vaisala, Tucson, 21–23 Apr 2008

Kitagawa N, Ohashi M, Ishikawa T (1990) Safety guide against lightning hazards. Res Lett Atmos Electr 10:37–44

Lengyel MM, Brooks HE, Holle RL et al (2005) Lightning casualties and their proximity to surrounding cloud-to-ground lightning. In: Preprints of the 14th symposium on education, American Meteorological Society, San Diego, 9–13 Jan 2005

López RE, Holle RL (1999) The distance between successive lightning flashes. NOAA Tech Memo NSSL-105, Natl Severe Storms Lab

Mainwaring L, Fricska G (2016) Lightning safety efforts in Canada 2009 to 2015. In: Preprints of the 6th international lightning meteorology conference, Vaisala, San Diego, 18–21 Apr 2016

Roeder WP (2007) The effectiveness of short notice outdoor lightning risk reduction and why it should not be taught. In: Preprints of the international conference on lightning and static electricity, Paris, 28–31 Aug 2007

Roeder WP (2009) Last minute outdoor lightning risk reduction-A method to estimate its effectiveness and comments on its utility in public education. In: Preprints of the 4th conference on the meteorological applications of lightning data, American Meteorological Society, Phoenix, 11–15 Jan 2009

Roeder WP (2014) Backcountry lightning risk reduction-Lightning crouch versus standing with feet together. In: Preprints of the 5th international lightning meteorology conference, Vaisala, Tucson, 20–21 Mar 2014

Roeder WP, Holle RL, Cooper MA et al (2012) Lessons learned in communicating lightning safety effectively. In: Preprints of the 4th international lightning meteorology conference, Vaisala, Broomfield, 4–5 Apr 2012

Sharma S (2012) Natural disasters; policy issues and mitigation strategies. Centre for Science and Technology of the Non-Aligned and Other Developing Countries, New Delhi. ISBN 978-81-7035-759-9

Trengove E, Lubasi FC (2015) Lightning science and lightning protection. Centre for Science and Technology of the Non-Aligned and Other Developing Countries, New Delhi. ISBN 978-93-5130-545-3

Uman MA (1986) All about lightning. Dover Press, New York. 167 pp

Vavrek J, Holle RL, Allsopp J (1993) Flash to bang. Earth Scientist 10:3–8

Walsh KM (1997) When thunder rolls: lightning on the football fields. Am Football Q 3:79

Walsh KM, Bennett B (1996) Making a case for lightning safety policy in intercollegiate athletics. In: Preprints of the international lightning detection conference, Global Atmospherics, Inc., Tucson, 4–6 Nov 1996

Walsh KM, Hanley MJ, Graner SJ et al (1997) A survey of lightning policy in selected division I colleges. J Athl Train 32:206–210

Walsh KM, Cooper MA, Holle R et al (2013) National Athletic Trainers' Association position statement: lightning safety for athletics and recreation. J Athl Train 48:258–270

Woodrum CC, Franklin D (2012) Using a lightning safety toolkit for outdoor venues. In: Preprints of the 4th international lightning meteorology conference, Vaisala, Broomfield, 4–5 Apr 2012

Zimmermann C, Cooper MA, Holle RL (2002) Lightning safety guidelines. Ann Emerg Med 39:660–665

Chapter 20
The Role of Lightning Warning Systems

Abstract To be effective, lightning warning systems must have several technical components that are discussed in this book. Even more important is that people must accept the risk as real, know what to do, and be willing to act in a way to save lives, including their own.

20.1 Lightning Risk

Prior chapters in this book discuss the ways in which lightning is a significant threat to human life as well as property and livestock. Part II described that lightning poses a risk resulting in 6000–24,000 fatalities a year globally. In developed nations, an estimate is that ten times as many people are injured, so between 60,000 and 240,000 people a year may be injured globally by lightning each year (Holle 2016). Of that population, a portion suffers short-term and long-term limitations and disabilities, both physical and mental, as identified in Part I. The toll on human life almost always extends far beyond the individual who is injured or killed to their families, friends, coworkers, and communities where they live. Preventing as many of these impacts as possible is the primary goal of this book.

Chapter 6 documents that awareness and acceptance of the lightning threat has had a positive effect in the developed regions of the world. Knowledge of when and where lightning is occurring and that it is a danger that can be avoided have been two of the many factors resulting in a lightning fatality rate that is more than an order of magnitude lower than a century ago in developed nations. The challenge is to bring these factors to developing countries to save lives.

M. A. Cooper, R. L. Holle, *Reducing Lightning Injuries Worldwide*,
Springer Natural Hazards, https://doi.org/10.1007/978-3-319-77563-0_20

20.2 Prerequisites to an Effective Lightning Warning System

As already stated, the three most important human factors in making a lightning warning system effective are the belief by the individual:

1. That they can be injured or killed by lightning
2. That they can do something about their personal safety when threatened by lightning
3. Their willingness to take action to avoid the threat

In fact, these may be the only factors that are important when a person is faced with lightning and must make the decision to take personal action. In other situations, lightning warning systems may involve more technical aspects as follows:

1. Belief that the lightning risk can be avoided
2. Accurate, testable lightning detection
3. Reliable forecasts and timely trustworthy warnings
4. Knowledge of action to be taken
5. Willingness and ability to take action
6. Availability of safe areas for evacuation or safe practices
7. Adequate time to reach safe areas
8. Willingness to stay in the safe area until an "all-clear" is issued or 30 min has passed since the last lightning was seen or thunder was heard

20.3 Lightning Safety Action Plan

The most important aspect of a lightning safety action plan is to have in mind when and where to reach a lightning-safe location. Other portions of preparing a plan are being aware of available weather forecasts for the area and planned time of activity, changing plans if thunderstorms are likely, and being actively aware of their surroundings for the danger of possible unexpected thunderstorms. For those with responsibility for others, such as teachers, coaches, and managers, the plan must allow enough time for everyone to reach safety.

Awareness and warnings are useless unless there are safe places to reach in time. There are two safe places from lightning (Chap. 16). One safe location is a substantial building that has conducting material within or around it to carry a direct or nearby lightning strike away from people inside. In developed countries, essentially all homes and workplaces have grounded wiring, plumbing, and often metal structural members that conduct a direct or nearby lightning strike around occupants and into the ground with minimal or no damage.

The other safe place is a fully enclosed metal-topped vehicle. While potentially frightening to anyone inside the vehicle if it is struck, such vehicles are excellent resources for lightning safety when no safe buildings are nearby. Lightning-safe buildings and vehicles need to be identified in advance in the workplace, home, and recreational situations. Each person must have a good idea of where lightning-safe places are located and how long it will take to reach them.

Technical warning systems involving sirens, horns, flashing lights, or other notification mechanisms may be available in certain venues such as sports stadia, golf courses, for those working at airports, and other situations. However, every person is responsible for their own personal safety and for those under their responsibility and must take action in threatening circumstances regardless of whether or not the lightning warning system seems to be functioning correctly.

Questions to Explore

How are lightning warnings distributed at your favorite sports or other outdoor venue? How much confidence do you have that the warnings are reliable? Do people heed them when issued? Do people go directly to a lightning-safe location? Do they stay there long enough?

20.4 Technical Lightning Detection and Warning

The lightning portion of the lifecycle of a thunderstorm is usually short-lived, since the lightning threat develops, grows, and ends in tens of minutes in most cases. Therefore, a warning system needs to be reliable and accessible on this short time scale (Chap. 21). There are established, objective statistical methods to measure the success of a warning system. First, the probability of detection (POD) needs to be high enough to be valuable. Typical successful POD numbers range from 0.75 to over 0.95 with several minutes' warning. The false alarm ratio (FAR) is equally important. Too many false warnings (those that do not result in lightning) quickly erode confidence in warnings that have been issued (Holle et al. 2016). For that reason, the warning system needs to be verified and tested in the area and time of year where the warning system is to be used, not in some ideal situation, since storm durations and sizes vary in differing meteorological situations.

Besides POD and FAR, there are the issues of how long in advance a warning should be made and how long to stay in a safe place before resuming outdoor activity. At the start of the lightning threat, at an airport, for example, 2 min is all that is often needed for ground workers to reach safety since they are usually very close to large lightning-safe buildings or vehicles (Holle et al. 2016). In other situations, it may take 10 min or more to reach safety. In that case, the POD and FAR will not be as good as for a short-time warning.

At the end of a storm, an individual working on a small project in their backyard may find that waiting 15 min after the last lightning and thunder is adequate since they can go back into the house very quickly if lightning returns.

For a large stadium or other gathering, a wait time of at least 30 min is highly undesirable to avoid too many back-and-forth movements of people. Such actions cause confusion and frustration and easily contribute to a lack of confidence in the lightning warnings (Walsh et al. 2013). When a lightning warning is issued at a sports event, the game is not only stopped, but everyone should be instructed to leave the field, viewing stands, and all surrounding exposed areas to go to previously specified safe locations (www.lightningsafety.noaa.gov/toolkits.shtml). Special plans must apply to programs occurring after dark. Lightning that is visible at night may not necessarily pose an immediate danger because it may often be seen on the horizon up to 30 miles away or more, much too far to pose an immediate threat.

20.5 Lightning Detection System Quality and Reliability

A lightning-vulnerable location where many people are at risk should have a proven, verified lightning detection system in place. Warnings based on low-cost, handheld, or untested detection systems can be highly misleading, dangerous, and untrustworthy (Chap. 14). Taking this approach to lightning warnings is potentially a symptom of failing to take the lightning threat seriously or of not doing adequate due diligence to assure accuracy and safety.

Finally, the chain of command for warnings must be robust. When a reliable warning is made, people need to proceed without question to a lightning-safe building, location, or vehicle that has been identified in advance. They should stay there until an unambiguous all-clear has been sounded using preset indications. Procedures and training of all personnel should be in place such that at night, and during weekends and holidays, the same system is objectively used and followed, since the warning procedure does not rely on a single person.

References

Holle RL (2016) A summary of recent national-scale lightning fatality studies. Wea Climate Soc 8:35–42

Holle RL, Demetriades NWS, Nag A (2016) Objective airport warnings over small areas using NLDN cloud and cloud-to-ground lightning data. Wea Forecast 31:1061–1069

Walsh KM, Cooper MA, Holle R et al (2013) National athletic trainers' association position statement: lightning safety for athletics and recreation. J Athletic Training 48:258–270

Chapter 21
How to Deliver the Message to Vulnerable Populations

Abstract The purpose of this book is to decrease deaths and injuries caused by lightning around the world by encouraging studies that gather and disseminate data to public safety planners, NGOs, policy makers, and others who are in positions to improve safety from lightning for all of their citizens. This chapter will cover methods for reaching the public, particularly the most vulnerable populations.

21.1 Introduction

As has been considered repeatedly in this book, the areas of the world and the populations that are most vulnerable to injury by lightning are those that also are least likely to have good communication systems to allow good public education (Cooper and Ab Kadir 2010). They are also the least likely to have safe areas to avoid injury and easy solutions to the lightning threat.

Public education and decreases in injuries are relatively easy to accomplish in developed countries because of the wealth and number of communication systems, excellent weather forecasting, and availability of safe areas. Messages such as "When Thunder Roars, Go Indoors" are easy to teach, and the biggest problem using it is translating it into catchy or memorable phrases in other languages.

Many methods have been used to deliver lightning safety messages in different venues. However, there are some specific limitations to some of these methods, which are also discussed in Chap. 20. Each of the approaches in Table 21.1 has advantages and disadvantages.

21.2 The Safety Message

The biggest problem in delivering lightning safety messages to the most vulnerable populations, at this time, is that there are no truly valid nor safe instructions that can be implemented after the populace receives them. Most are, at best, partial measures. The primary message is to seek either a fully enclosed metal-topped vehicle

M. A. Cooper, R. L. Holle, *Reducing Lightning Injuries Worldwide*,
Springer Natural Hazards, https://doi.org/10.1007/978-3-319-77563-0_21

Table 21.1 Methods that have been used to deliver lightning safety messages, listed alphabetically

Comic strips
Distribution of leaflets, booklets, posters
Education of coaches and teachers
Exhibitions and trade shows
International Lightning Safety Day, 28 June
Internet, cell phones, Twitter, blogs, and other social media
Mobile apps
National-level quiz programs
Public and community education
Roadshows and billboards
School curricula, children's games, puzzles
Science documentaries on lightning and lightning injuries
Seminars, workshops, and conferences
Street theater, folk songs, dances, storytelling
Television, radio, print media
Use of special spokespersons such as media, lightning survivors, meteorologists, broadcasters, well-known sports figures, and others with an interest in presenting injury prevention messages
www.lightningsafety.noaa.gov

or a substantial building, neither of which may be readily available to vulnerable populations. There is often no safe shelter available in rural, mountainous, or wilderness areas where a large proportion of the population lives in many countries.

A study in Colombia noted that while the majority of deaths were to those living in the cities due simply to their large populations, the highest proportion of deaths occurred in rural areas where no safe shelter was available (Navarrete-Aldana et al. 2014). It was concluded that different methods should be used to deliver safety messages in these two different settings. Yet, no research has been done on what this message should be.

Research Questions

What safety messages can we develop for the people who live in sub-Saharan Africa, the mountains of Colombia, or the agricultural fields in Bangladesh that are valid, reliable, and easy to implement? Should these be tested before they are disseminated? How?

21.3 DO Messages Versus DON'T Messages

Many sets of safety messages have concentrated on giving a large number of DON'T messages instead of giving simple what to DO messages (Table 19.3; Figs. 21.1 and 21.2). This approach results in a series of confusing issue-specific recommendations. It is far more effective to give one or two simple messages that people can remember

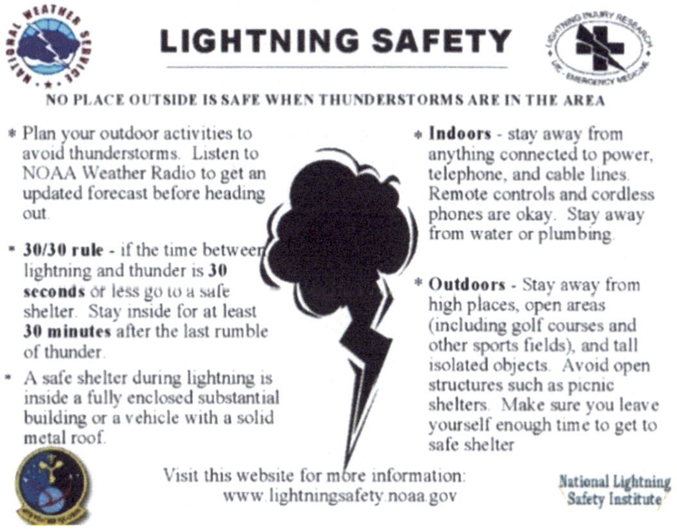

Fig. 21.1 Lightning safety magnet used by US Lightning Safety Week in 2002–2006 with a long list of DON'T messages

Fig. 21.2 Lightning safety magnet used by National Lightning Safety Week from about 2007–2012 with a simple DO message

in times of fear and stress than to give a long list of what they should not do, which they are unlikely to remember (Fig. 21.1). However, as noted, no effective, implementable messages have been developed for the most vulnerable populations.

21.4 The Weather Message: Forecasting

In many developing countries, high-quality, real-time weather forecasting is limited to servicing the international airports. While droughts, floods, hail, and other weather risks are well known to affect food production and other important economic functions, the resources to gather the appropriate data and deliver forecasts for the rest of the country may be lacking.

In developed countries, weather forecasting is often considered a government service, not only providing day-to-day forecasts but also contributing to public safety with severe weather warnings. There are often multiple commercial sources of weather forecasts accessible by television, radio, internet, mobile phone, and other venues. These systems, many supported indirectly by advertising, compete for viewership. The public is free to judge the reliability and quality of the forecast and visual presentation. The frequency of accessing the sources dictates which are successful and will continue to be in demand. Warning system apps may be free, and more intensive monitoring for specific venues may be available by subscription. Some sports and entertainment venues even install their own warning and monitoring systems.

In developing countries, not only is the forecast availability and quality variable, but there may be little opportunity for free market competition. Laws in some countries prohibit the dissemination of weather information by sources outside the national meteorological service, some with significant financial and prison penalties. Other countries may mandate that their meteorological agencies adopt a business model, requiring users to pay for weather information, including paid subscriptions for cell phone apps. While the cost may not seem significant to those in cities or in developed countries, it can be prohibitively expensive for those in most need of warnings in developing countries: the subsistence farmer, fishermen, and other rural people.

21.5 Other Barriers to Messaging

21.5.1 Language

In countries with many languages and dialects, appropriate translation and dissemination to vulnerable populations may be difficult to achieve.

21.5.2 Literacy

Gomes et al. (2006b) proposed a model for lightning awareness in third world countries based on literacy levels, suggesting that in areas where the literacy is above 90%, the internet can be used as one of the ways of creating lightning safety awareness. However, the definition of literacy varies with country. For instance, literacy in South Africa is defined as completion of the 7th grade, although it is recognized, given the quality of schools, that many 7th graders cannot read (Trengove and Jandrell 2012). In Uganda, only half of those surveyed had completed primary school (Tushemereirwe et al. 2017).

21.5.3 Electricity, Internet, and Cell Tower Availability

The availability of reliable power hampers not only economic development but also the availability of internet and mobile phone service.

21.5.4 Isolation

The most vulnerable populations often fall into two classes:

1. Refugees
2. Rural populations who may be dispersed, nomadic, or itinerant and are hard to reach in large numbers

21.6 Methods for Delivering Lightning Safety Information

21.6.1 Television, Radio, and Print Media

Television and radio are probably the most effective means of delivering safety messages and other information. They require no literacy; the broadcasts are usually free and often in the prevailing language in the area where the safety message is needed. The disadvantage is the unavailability of electricity to power them and the cost of television sets. Radios are much less expensive and more easily battery powered. Everyone recognizes that print media is dying, at least paper versions. Effective dissemination is dependent on the popularity of the print venue, literacy, and where the message is located in the publication. If it is buried in the middle of the publication, it is unlikely to be seen.

21.6.2 School Curricula, Children's Games, Puzzles

As propagandists have known for centuries, the best and often quickest way to educate a population is to work with the children. While difficult to achieve, it is probably optimal to have lightning safety, as well as other safety lessons, incorporated into the national curricula rather than trying to give special sessions at individual schools. Many free resources already exist at www.lightningsafety.noaa.gov, and writers of the site are willing to work with people from other countries to translate, modify, and personalize materials to show children and situations that are from those countries. Parents and schools should have a particular interest in this due to the large numbers of children injured by lightning at schools (Holle and Cooper 2016).

21.6.3 Public and Community Education

Multiple methods can be used depending on literacy, population concentration, and other factors (Gomes et al. 2006a, b, 2012). Some methods, such as street theater, storytelling, and singing, often reach only small numbers due to the dispersion of rural villages (Table 21.1). Nevertheless, it may be effective in areas of low literacy. As noted before, the use of print materials may be good for teachers but less useful depending on literacy and the multiple languages in which they might need to be produced.

21.6.4 International Lightning Safety Day, June 28

The African Centres for Lightning and Electromagnetics Network (ACLENet) discovered that many deaths in African newspaper reports during the 2000s were to school children (https://aclenet.org/projects/save-a-life-in-africa). One incident alone involved the deaths of 18 children and injuries requiring hospitalization of 38 more on June 28, 2011 (http://www.telegraph.co.uk/news/weather/8606238/Lightning-strike-kills-18-children-in-Uganda.html). Consequently, a resolution for an International Lightning Safety Day to commemorate this tragic day of June 28 was passed at ACLENet's Second Symposium and adopted by the representatives of the 17 countries attending. Several countries have begun holding lightning safety activities on these days.

21.6.5 Use of Social Media, Internet, and Mobile Phones

There has been a huge explosion of mobile phone and social media usage throughout the developing world. Mobile phones are used for accessing health messages on HIV, child care, and other areas, for information on crops and animal husbandry, for checking market and fuel prices, and for transferring money (Southwood 2009; Wasserman 2011; Chiumbu 2012; Trengove and Jandrell 2012). The Arab Spring and daily political protests in Venezuela were driven by text messages, Snapchat, and other social media, because all other media sources have been shut down (Gire 2017; Lopez 2017; Wilson 2014). In partnership with Airtel, Human Networks International's 3-2-1 free service started in Madagascar in 2010 and was generating over 250,000 calls per month covering 350 messages by 2015 (HNI.org). These are all examples of person-accessed messages, not broadcast warnings.

Trengove and Jandrell (2012) posited that using mobile phone texting to issue lightning warnings and education would have the following impacts:

1. Reach a large number of people
2. Reach rural people
3. Bridge the digital divide by providing the same service to rich and poor
4. Could use existing mobile telephone infrastructure
5. Could geographically target lightning warning messages

Some of these hypotheses have worked out, while others are more limited. Unfortunately, despite decreasing costs in many countries, smartphones remain prohibitively expensive for the poor and rural populations of most countries. Tushemereirwe et al. (2017) noted that while 92% of Ugandans surveyed had mobile phones, only 4% had smartphones. Cell phone penetration in 50 other countries, surveyed by Newzoo's Global Mobile Market Report (2017), is shown in Table 21.2. Similar data from Tech in Asia is shown in Fig. 21.3 for other countries in Africa and the Asia-Pacific region.

Floods, drought, and severe storms tend to disproportionately affect women since they are more commonly responsible for farm labor, food security, and household management in developing countries. The majority of farmers in Uganda are women, yet gender disparities limit their access to information on which to base decisions and adjust to climate shocks (Kyazze and Kristjanson 2011). This is caused by women's restricted access to technology and communication channels, lower education levels, and culturally defined roles in household chores such as raising children and cooking. In Africa, women are 23% less likely to own a mobile phone than are men (GSMA 2013). Financial barriers such as the inability to pay fees or even to own a mobile phone or radio can leave them uninformed of weather-related impacts.

Airtime can be expensive and unreliable. Electricity may only be available erratically, leading to mismatches between a person or home's allotted electricity window and the internet provider's window. As in most countries, there are multiple airtime providers so that not all of the population would likely be covered, and it would require funding by each company. For the HNI-Airtel partnership mentioned earlier, HNI, through private funding, provides the translated messages and Airtel provides a monthly allowance of free calls to promote customer loyalty. Not all telephone services are this benevolent.

21.7 Community Warning Systems

Over and above giving lightning safety messages to vulnerable populations, an additional level of safety would be for severe weather and lightning warnings to be issued or "pushed" through some sort of warning mechanism, similar to that used at airports in Australia (Potts 2009), Hong Kong (Li and Lau 2008), and the United States (Holle et al. 2016). These warning systems are typically operated by individual airlines using lights, sound alerts, electronic messaging, and other methods to inform ground crews in a noisy environment of the lightning danger within specific distances and times. The distance and time criteria are evaluated for safety versus efficiency since minutes of downtime due to airport ramp closures can have large downstream economic consequences (Steiner et al. 2013). If it were economically feasible, a school system or a small village could have warnings that triggered a flashing light or siren system, warning those to take shelter in safe areas. This would obviate the need for literacy but would reach only those in the immediate area within sight or sound of the warning system.

Where this type of system is not possible or affordable, Tushemereirwe et al. (2017) attempted to determine what Lake Victoria fishermen thought they needed for a warning system. On Lake Victoria, it is estimated that 5000 fishermen lose

Table 21.2 Smartphone penetration by population and country in April 2017

Rank	Country	Population	Smartphone penetration (%)
1	United Arab Emirates	9,398,000	80.6
2	Sweden	9,921,000	72.2
3	Switzerland	8,454,000	71.7
4	South Korea	50,705,000	71.5
5	Taiwan	23,564,000	70.4
6	Canada	36,626,000	69.8
7	United States	326,474,000	69.3
8	Netherlands	17,033,000	68.8
9	Germany	80,636,000	68.8
10	United Kingdom	65,511,000	68.6
11	Australia	24,642,000	67.7
12	Belgium	11,444,000	67.3
13	Spain	46,070,000	66.8
14	Azerbaijan	9,974,000	66.4
15	Italy	59,798,000	65.8
16	France	64,939,000	65.3
17	Saudi Arabia	32,743,000	65.2
18	Portugal	10,265,000	65.0
19	Czech Republic	10,555,000	64.8
20	Malaysia	31,164,000	64.1
21	Poland	38,564,000	63.4
22	Greece	10,893,000	59.5
23	Chile	18,313,000	56.0
24	Romania	19,238,000	56.0
25	Russia	143,375,000	54.7
26	China	1,388,233,000	51.7
27	Japan	126,045000	50.1
28	Turkey	80,418,000	49.8
29	Argentina	44,272,000	48.2
30	Mexico	130,223,000	40.7
31	Thailand	68,298,000	40.5
32	Kazakhstan	18,064,000	39.2
33	Brazil	211,243,000	37.7
34	Iran	80,946,000	37.1
35	Venezuela	31,926,000	36.2
36	South Africa	55,436,000	36.2
37	Peru	32,166,000	36.0
38	Colombia	49,068,000	35.4
39	Morocco	35,241,000	33.4
40	Algeria	41,064,000	32.4
41	Egypt	95,215,000	30.4
42	Vietnam	95,415,000	26.4

(continued)

Table 21.2 (continued)

Rank	Country	Population	Smartphone penetration (%)
43	Ukraine	44,405,000	23.5
44	Philippines	103,797,000	23.3
45	India	1,342,513,000	22.4
46	Indonesia	263,510,000	20.7
47	Iraq	38,654,000	19.9
48	Nigeria	191,836,000	14.8
49	Pakistan	196,744,000	9.2
50	Bangladesh	164,828,000	5.2

Source: https://en.wikipedia.org/wiki/List_of_countries_by_smartphone_penetration

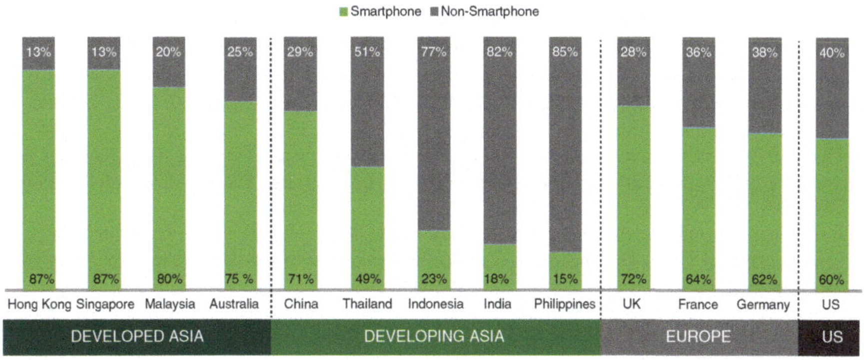

Fig. 21.3 Smartphone penetration for the Asia-Pacific region. Credit: "User Happiness, Tech in Asia." Source: https://www.techinasia.com/nielsen-report-smartphone-adoption-gap-asia-pacific

their lives to severe weather every year. They prefer severe weather warnings that come by (1) text, (2) vibration, or (3) color-coding and a loud noise for when they were busy with their nets or for those with lower literacy.

Of course, targeted warnings in either of these settings would depend on GPS coordinates, reliable and consistent internet or cell phone availability, how long thunderstorms last in this region, and the speed of movement of a thunderstorm. Additional factors include the quality and timeliness of the forecast and the willingness and funding of the meteorological service to implement such a system.

References

Chiumbu S (2012) Exploring mobile phone practices in social movements in South Africa – the Western Cape Anti-eviction Campaign. Afr Identities 10(2). https://doi.org/10.1080/14725843.2012.657863

Cooper MA, Ab Kadir MZA (2010) Lightning injury continues to be a public health threat internationally. In: Preprints of the 21st international lightning detection conference, Vaisala, Orlando, 21–22 Apr 2010

Gire S (2017) The role of social media in the Arab Spring. https://sites.stedwards.edu/pangaea/the-role-of-social-media-in-the-arab-spring/. Accessed 8 Sept 2017

Gomes C, Ahmed M, Abeysinghe KR et al (2006a) Lightning accidents and awareness in South Asia: experience in Sri Lanka and Bangladesh. In: Proceedings of the 28th international conference on lightning protection, Kanasawa, 18–22 Sept 2006

Gomes C, Kithil R, Ahmed M (2006b) Developing a lightning awareness program model for third world based on American-South Asian experience. In: Preprints of the 28th international conference on lightning protection, Kanazawa, 18–22 Sept 2006

Gomes C, Chowdhury AM, Ahmed M et al (2012) .South Asian Lightning Awareness Program (SALAP) Sri Lanka, Bangladesh and Bhutan. pdf.usaid.gov/pdf_docs/PNADD530.pdf. Accessed 8 Sept 2017

Groupe Speciale Mobile Association (2013) Women & mobile: a study on the mobile phone gender gap in low and middle-income countries. 6 pp

Holle RL, Cooper MA (2016) Lightning-caused deaths and injuries at schools. In: Preprints of the 33rd international conference on lightning protection, Estoril, 25–30 Sept 2016

Holle RL, Demetriades NWS, Nag A (2016) Objective airport warnings over small areas using NLDN cloud and cloud-to-ground lightning data. Wea Forecast 31:1061–1069

Kyazze F, Kristjanson P (2011) Summary of baseline household survey results: Rakai district, south central Uganda. November 2011. CGIAR Research Program on Climate Change, Agriculture and Food Security (CCAFS)

Li PW, Lau DS (2008) Development of a lightning nowcasting system for Hong Kong International Airport. In: Preprints of the 13th conference on aviation, range and aerospace meteorology, American Meteorological Society, New Orleans, 20–24 Jan 2008

Lopez V (2017) .On the frontline of Venezuela's punishing protests. https://www.theguardian.com/world/2017/may/25/venezuela-protests-riots-frontline-caracas-nicolas-maduro. Accessed 30 Oct 2017

Navarrete-Aldana N, Cooper MA, Holle RL (2014) Lightning fatalities in Colombia from 2000 to 2009. Nat Hazards 74:1349–1362

NewZoo (2017) Top 50. countries by smartphone users and penetration. https://newzoo.com/insights/rankings/top-50-countries-by-smartphone-penetration-and-users/. Accessed 27 July 2017

Potts R (2009) A thunderstorm and lightning alert service for airport operations. In: Preprints of the 4th conference on the meteorological applications of lightning data, American Meteorological Society, Phoenix, 11–15 Jan 2009

Steiner M, Deierling W, Bass R (2013) Balancing safety and efficiency of airport operations under lightning threats: a look inside the ramp closure decision-making process. J Air Traffic Control 55:16–23

Trengove E, Jandrell IR (2012) Leveraging a mobile culture for lightning awareness: the African context. In: Preprints of the 31st international conference on lightning protection, Vienna, 2–7 Sept 2012

Tushemereirwe R, Cooper MA, Tuhebwe D et al (2017) The most effective methods for delivering severe weather early warnings to fishermen on Lake Victoria. PLOS Current Disasters, February 22, 16 pp, doi:https://doi.org/10.1371/currents.dis.d645f658cf20bc4a23499be913f1cbe1

Southwood R (2009) Less walk more talk: how Celtel and the mobile phone changed Africa. Wiley: New York

Wasserman H (2011) Mobile phones, popular media, and everyday African democracy: transmissions and transgressions. Int J Media Cult 9:146–158

Wilson P (2014) .Social media key for Venezuelan protesters. https://www.usatoday.com/story/news/world/2014/02/19/venezuela-uprising-protests/5606899/. Accessed 8 Sept 2017

Chapter 22
How to Write a Grant to Perform Studies

Abstract Grant writing can seem to be an overwhelming and complex affair. However, it can be broken down into smaller steps such as finding grants, learning to read and screen grant requests for proposals, recruiting collaborators, preparing and submitting a letter of intent, writing an ongoing project description that can be inserted into a proposal, writing the proposal when requested, and preparing the budget. This chapter will describe how to obtain grants, whether as an individual trying to fund a research project or a nongovernmental organization, not-for-profit, or other type of organization wishing to conduct research or do service projects and education.

22.1 Introduction

The good news: Many very useful and high-quality studies can be done with little or no funds. Many, if not most, of the population studies, mechanisms of injury, and other papers by the two authors of this book have been done on our own time and were self-funded. Some research was indirectly supported by salaries that we drew from our employment (University of Illinois at Chicago and National Severe Storms Laboratory) when performance expectations included research and publications. We have also developed a network of colleagues and resources that we can call on for assistance with details and background or have the support of our current employers (Vaisala) for specific parts of our research such as lightning density maps. In general, people enjoy being asked for their assistance and advice and are especially happy if they know their support will save lives.

Most of the projects that we have mentored in other countries, which have resulted in publications and cross-connections for the authors with colleagues with similar interests, have also been done by individuals on their own time through dedicated efforts. If the project is an educational program or service project, bringing this network together may gather people with a wide range of complementary talents that can be used to make the study or project worthwhile and potentially worthy of journal publication. For example, the Lightning Safety Group in the United States is a collection of volunteers with various interests in lightning across the country

M. A. Cooper, R. L. Holle, *Reducing Lightning Injuries Worldwide*,
Springer Natural Hazards, https://doi.org/10.1007/978-3-319-77563-0_22

who rarely meet directly (Chap. 18). Nevertheless, they have been able to maintain and grow connections and contribute resources to the most comprehensive lightning safety website in the world at www.lightningsafety.noaa.gov.

22.2 Learning How to Write a Grant

Ideally, one learns to write a grant from mentors, an adviser or graduate supervisor, who are experienced in grant writing in your area of interest. Unfortunately, not all of us have those resources available.

There are multiple organizations that serve nongovernmental organizations (NGO), not-for-profit (NFP), and other nonprofit organizations, and many have excellent, and often free, webinars on grant writing and other fundraising methods. At the time of this writing, some of these include Firespring, FundsforNGOs, TechSoup, Nonprofit Hub, CharityHowTo, and GrantStation. Googling will often find others, and once your organization is registered with one, the others may connect with your address and begin mailing you as well.

22.3 Preparing a Grant Proposal

22.3.1 Finding Research Funds and Grants

With the exception of an individual student's project, which may be funded from their adviser's grant funding, the vast majority of grants are written to fit within a specific request for proposal from a grant-making foundation, government body, or other funder. A grant written as a wish with no specific funder in mind is likely to be a waste of time that could be more valuably spent searching for appropriate grants.

In order to perform a complete project, it may be necessary to have several grants for different pieces of the project. Conversely, for large projects, it may be necessary to include other agencies and NGOs as collaborators to provide the talents, labor, and expertise needed to assemble a good team that can convincingly carry out the proposed project. Currently, grant makers are shifting to results-based proposals and often require collaboration of several organizations in order to show reach and availability of requisite skills to accomplish the objectives of the grant. Theory of Change has become one of the current touted methodologies for some funders.

An undergraduate or graduate student may be able to tap into funds that their professor/adviser already has to support his/her work. Generally, a student needs funds for a short but immediate amount of time, and their adviser should be their first source for money, in-kind support, and advice. A faculty member may need financial support for travel or other costs to perform studies. Both groups may be able to tap into departmental, dean's, or university discretionary funds or small grant programs for small amounts. An NGO, usually with longer timelines, may have more time to search for grants, supporting their operational overhead with overlaps from other grants or monies in the meantime.

There are a number of ways to find grants. University faculty should always look within their department or home institution, as many schools have grant offices of various sizes, qualities, and priorities. Grants are very important for universities to thrive and are not a burden but an opportunity. Many universities offer courses in grant writing or have personnel who can help the candidate find the keywords necessary to narrow a search, teach the candidate how to do the search, or do it for them. However, the best person to do a grant search is usually the one with the most to gain and has the best knowledge base in the area. There are groups, programs, and organizations that assemble grant proposals with subscriptions. Two of these are GrantStation (mostly United States) and FundsforNGOs, but there are many additional resources.

Networking is very important. Others in the same interest area may know grant sourcing/searching programs unknown to you. If one that you find doesn't exactly fit your goals, it is easy to forward it to contacts and organizations who might find them applicable, engendering goodwill. Grant referrals or collaborative opportunities that you can use may come back from those in your network where you sent calls for proposals.

All of this takes time and energy. It is not glamorous, but it may well be essential for your research or for the viability of your organization.

22.3.2 Formulating the Study Question: Doing Your Homework and Being Prepared

Before writing a grant, it is important to:

- Hone your research into a testable hypothesis
- Define the methodology that will define how and what data will be collected
- Identify a population or issue to be studied, laboratory to conduct the research if appropriate, essential collaborators, data analysts, and other resources necessary to carry out the project
- Prepare a budget

For NGOs, the mission and vision statements should lead to goals/objectives. These can be divided into more specific research and service projects. Background data, references, statistics, and other supporting data should be assembled to write a description for each project. As projects are finished and new ones begun, the supporting documents can be updated and expanded to support ongoing or more advanced programs that build on those already completed and to demonstrate a track record.

Any letter of intent, letter of interest, expression of interest, proposal, position paper, or grant application, whether funded or not, should become part of a database whose contents can be used in future grant applications as well as serving as documentation that the organization is capable of carrying projects to completion. It is highly recommended to look through grants that were successfully funded to organizations similar to yours by a specific foundation or agency. These are often available on the web. Of course, applications will always need to be tailored to the specific grant requirements.

22.3.3 Determining If a Grant Is a Good Fit for Your Project or Organization

There are two ways of looking at a project: from the view of those doing the project and from the view of the grant maker. The closer the match between these two views, the more likely the grant will be successful. In every case, the proposal must align very closely with the goals of the granting foundation, or it will not be considered. However, making impossible promises to fit a request for proposal and going too far afield from your organization or research project's goals and objectives will not lead to a good end.

22.3.4 Setting Criteria for Your Project/Study

The following are questions to be asked as the letter of intent and subsequent proposal is prepared:

- What do you want to accomplish or study?
- What methodology will you use?
- How long will the study last?
- What are the necessary equipment, supplies, personnel, materials, or other components?
- How many people, animals, and schools will be needed for the research?
- What geographic area will be involved?
- What are the other aspects of the project/study that are essential to bring it to completion?
- Which of all of these questions are flexible and which are not?
- What difficulties and risks do you foresee?
- What costs are involved?
- What outcomes are to be completed within the allotted time of a grant?
- If funding for essential portions of your project are excluded by the grant maker, where will you find funding for these?

22.3.5 Reading the Request for Proposal (RFP)

Whenever an announcement, grant link, or request for proposal (RFP) is found that may be a potential fit, the first step is reading the RFP *carefully* to see if you or your organization qualifies, is large or small enough, has the required track record and expertise, and works in the funded geographic area and other important requirements. Part of reading the proposal is also evaluating if it fits the goals of your project or organization, as in Table 22.1.

Most of the points in Table 22.1 are self-explanatory. However, finding the grant that is the best fit for the project or topic you propose may not be straightforward. You need as close a match as possible, but you will never find a perfect fit. Sometimes your question can be answered within the larger framework of the RFP vision. For instance, an RFP may be tailored to climate change and resilience in Africa, making the people more able to withstand climate and weather threats. In this example, the following are some directions that may be taken:

1. If your project is to determine the effect of lightning injuries over the past decade, a grant to survey several different weather threats to farmers, including lightning, may be more readily fundable since it covers a broader topic area for the RFP to cover.
2. If instead, you are interested in delivering public education to prevent injuries and enable villagers to carry out their normal activities without as much fear from lightning, doing a baseline study to assess injuries, beliefs about lightning, and other factors will help to develop the educational programs, which can be delivered to the schools, parents, or local population and then assessed for its effectiveness.
3. If you are interested in lightning detection, you may be able to write the grant to use lightning as a proxy for severe storms that destroy crops. Including collaborators, advisors, or mentors who are familiar with farming in the region may make your proposal more presentable.

Although it may be determined that everything else fits, the last two items in Table 22.1 may eliminate an application. For NGOs, sometimes pre-proposal work

Table 22.1 Considerations in reading a request for proposal

Work or research topic to be covered
Time period to be covered
Other exclusion criteria
What will be funded – or not funded
Need for partner documents, track record
Other requirements
Submission deadline
Amount of pre-proposal work that must be completed

completed for other projects can be modified and used again. Existing support letters, signed documents, and other materials will need to be updated with the new date, title of the project, addressee, and grant maker along with permission of the collaborators to use for the new project.

22.3.6 Ongoing or Multiple Projects

More complex organizations may find it useful to assemble a grant decision matrix, grants calendar, and grants pipeline. A grant decision matrix is a scoring tool that can help organizations triage which RFPs are worth pursuing with an application. Table 22.2 provides some considerations that will help in developing a letter of intent and possible subsequent proposal.

The criteria, rating system, multipliers, and decision score action distribution such as the example in Table 22.3 are all for the use of the NGO, and there are no right answers. The categories can be chosen to suit the NGO, the grant writing team, the NGO's Board of Directors, or other appropriate agents, and should probably be followed and revisited periodically, particularly if they are not found to be good predictors of success. However, this self-rating needs to be honest so that the evaluation is not unduly optimistic or pessimistic. Some criteria, such as eligibility, may be considered "absolutes" and inappropriate to put in the matrix.

Table 22.2 Hypothetical grant decision matrix

Decision criteria to be rated	Excellent			Poor	Multiplier	Score
How well does it fit the NGO's mission?	0	1	2	3	3	x
How well does it serve target population?	0	1	2	3	3	x
Eligibility	0	1	2	3	5	x
Timing of letter of intent	0	1	2	3	3	x
Timing of proposal if invited to submit	0	1	2	3	2	x
Available/qualified staff	0	1	2	3	3	x
Will grant funds be adequate?	0	1	2	3	3	x
Matching funds required?	0	1	2	3	3	x
Administration costs allowed	0	1	2	3	3	x
Other criteria	0	1	2	3	2	x
Total						X

Table 22.3 Action based on decision matrix score (fitted to your NGO)

Score	Decision
0–30	Apply
30–50	Refer to director for further discussion and consideration
50–75	Reject

A grants calendar can be as simple or as complex as needed and can take on layers depending on how many letters of interest and grant submissions are being prepared. For simple grants, a calendar with task completion, editing, assembly, and submission dates may be all that is necessary. For large projects, where NGOs are writing many letters of intent and subsequent grants at the same time, each letter and grant will need its own work matrix or Gantt chart which may drive the work of several team members and contributors. An overall summary chart will be essential to make sure that everyone is doing their portion and that no one person is overwhelmed with too many project pieces coming due simultaneously.

22.4 Other Considerations

This chapter cannot begin to be a complete tutorial in grant writing, but here are a few other pointers:

22.4.1 Definitions

Make sure you know the meaning of the terms the grant maker is using. Do not assume a common or older definition applies. This is true for administrative factors such as bookkeeping, NGO registration, and audits for technical scientific or analytic terms and especially for current trigger words such as climate change, resilience, sustainability, etc. Trigger words are especially important to repeat throughout the proposal, although they should be used accurately and appropriately.

22.4.2 Methodology

If a certain methodology is required for the application or evaluation of the outcomes of your project, do your best to become familiar with how it is applied. It may be useful for the next time you write a grant.

22.4.3 Content Readers

Depending on the complexity of the proposal, it may be desirable to have a content area reader read the proposal for technical accuracy, statistics, and methodology. Someone who is familiar with the NGO's mission and projects should read the grant for flow and agreement of the points presented, to decrease repetition and to suggest visual interest (pictures, graphs, colors). Before the final copy is submitted, another reader should read the proposal for grammar, spelling, and organization before the final checklist of proposal components is assembled and rechecked. If funds are available, use a professional grant reviewer.

22.4.4 Relationships

When possible, build a connection with a person within an organization that may be related to your needs for funding. Respectfully request a grants manager to provide information on when the next round of grant solicitations is likely to be issued and the general terms that may be included. Typically, a grants manager relies on a committee to decide on the merits of submitted letters of intent as well as the proposals that are submitted. Some of the same committee members may serve on panels for different organizations, so be sure that a quality letter of interest is provided that fits the grant guidelines. In addition, the grants manager may be able to determine quickly that a letter of inquiry is not relevant for that organization but may know of others in the grant community that are a better match.

22.4.5 Timeliness and Word Count

Late submissions or those exceeding the designated word count may be immediately rejected. Submit a letter of intent or proposal several days before the due date to be sure that any communication issues, local holidays, and weekends are taken into account. For hard copies that are mailed, a trackable mail system should be used. For those submitted electronically, a delivery receipt and a read receipt can be requested so that the grant writer knows their work has been safely delivered.

Index

M. A. Cooper, R. L. Holle, *Reducing Lightning Injuries Worldwide*, Springer Natural Hazards, https://doi.org/10.1007/978-3-319-77563-0

The manufacturer's authorised representative in the EU is Springer
Nature Customer Service Centre GmbH, Europaplatz 3, 69115 Heidelberg,
Germany. If you have any concerns regarding our products, please
contact ProductSafety@springernature.com

Printed and bound by CPI Group (UK) Ltd, Croydon, CR0 4YY

05/10/2026

02228139-0004